"What a terrific book about a great scient live in, and the personal and technical challenges he faced. Luke Timmerman has done a remarkable job, through numerous interviews, of capturing all aspects of Lee Hood's amazing life story."

Robert Langer, Institute Professor at MIT, winner of the National Medal of Science and National Medal of Technology

"A thorough biography [that] never shies away from the contradictions of this forceful personality. ... Hood deserves to be a household name."

Kirkus Reviews (starred review)

"I couldn't put this book down."

Atul Butte, director, Institute for Computational Health Sciences, University of California-San Francisco

"Timmerman captures the glory and clay feet of a great scientist, like Jim Watson did for himself in *The Double Helix*."

Phillip A. Sharp, Institute Professor at MIT, Nobel Prize-winning biologist

"A great book, with great insight and reporting. If there is one lesson I hope we all learn from the book, it is that Lee saw a future no one else saw, and the many fruits that come from his vision include the lives saved of patients who don't even know his name."

Irving Weissman, director of the Stanford Institute for Stem Cell Biology and Regenerative Medicine

"I devoured *Hood*. This book provides an intimate window into one person's life, contributions, personality, and relationships. As someone who has dedicated a career to helping speed up drug discovery, and remove barriers that get in the way of scientific progress, I found this book to be profoundly illuminating."

Margaret Anderson, executive director, FasterCures, a Center of the Milken Institute

"A fascinating human account of the life of a person who truly has changed the world, persevering in the face of consistent establishment skepticism to achieve his visions."

Ed Lazowska, Bill & Melinda Gates chair of computer science and engineering, University of Washington

"Scientific discovery and technological innovation don't just happen; they are the product of relentless personalities charging through inertia to realize seemingly quixotic ambitions. *Hood* meticulously and entertainingly chronicles Lee Hood's path to greatness, and in the process offers a rare glimpse into the sausage making of our most recent scientific revolution. Timmerman's prose hooked me on the first page, and held me till the last."

D.A. Wallach, partner, Inevitable Ventures; recording artist

"More than any scientist I know, Lee Hood can see a picture of the future that hasn't been invented yet and figure out how to get there."

Robert T. Nelsen, managing director, Arch Venture Partners

"Lee Hood is an extraordinary man of many ideas, many talents, and many slide projectors. Luke Timmerman fully captures this Willy Wonka of molecular biology and all of his complications in an utterly riveting narrative."

Misha Angrist, associate professor of the practice, Social Science Research Institute, Duke University; editor-in-chief, *Genome* magazine

"A compelling, deeply human read about one of our country's most inventive scientists—and an unobstructed view into the egos and drama of modern high-stakes research."

David A. Shaywitz, chief medical officer, DNAnexus

"A tour de force of biography, the kind you rarely see done on a living scientist. Anyone curious about the biotech revolution we're living in today should read this."

Thong Le, CEO, Accelerator Corp.

"*Hood* tells the story of a legendary scientist and very human man, but if you read carefully, it also tells the story of personalized medicine."

Michael Pellini, chairman of the board, Foundation Medicine, Inc.

"A revealing story that should resonate beyond the biotech world."

Todd Bishop, co-founder, GeekWire

"An engrossing look at the man, his times, the technology he brought into being, and the implications for humanity."

San Diego Union-Tribune

"A thrilling ride through the life of a visionary biologist."

Mendelspod

"A page-turner."

Bio-IT World

HOOD

New directions in science
are launched by new tools
much more often than by new concepts.

The effect of a concept-driven revolution
is to explain old things in new ways.
The effect of a tool-driven revolution
is to discover new things
that have to be explained.

Freeman Dyson,
Imagined Worlds

Table of Contents

Foreword

Lee Hood is a legend. He has produced a raft of scientific accomplishments, and that is only a part of the story. He has a perspective that gives each of his talks a distinctive quality: he is always looking ahead. Other visiting speakers may come to a university to display what they have accomplished. For Lee, that is old news. He comes to tell his audiences what he sees happening in the future. He regales us with how he is helping to create that future, giving hints of what is to come. He has been doing this for decades and I have heard various renditions of his futurism at scientific meetings, at university seminars, and in living rooms in Montana. The extraordinary thing about Lee's predictions is how often they come to pass.

But even if his predictions are not fulfilled—and he usually promises more than is delivered—his talks have one remarkable consequence: the young people who come are transfixed. They realize that the enterprise they are preparing themselves for is a huge one, with wide ramifications. They realize that the small advances they struggle to make daily are part of a greater endeavor to understand living organisms, to deal with the imperfections of inheritance, and to counter disease. They see in Lee someone who has encompassed this larger vision and is consumed by making it the reality.

Lee studied and worked at Caltech, the university where I have been for the past twenty years. He had left years before I arrived, but his outsize personality and his drive left a huge legacy at Caltech. People remembered his many ideas, his large and ever-increasing laboratory, his leadership and his friendship. He had started at Caltech as an undergraduate, went off to medical school at Johns Hopkins, came back as a graduate student, went off to the NIH for a short stint, and then returned to Caltech as a faculty member for more than twenty years. His impact on immunology was enormous. For his graduate work he filled out the thoughts of his mentor, Bill Dreyer, and they presented strong evidence that antibody genes had both variable and constant regions joined into one molecule. How that might happen was a mystery solved by Susumu Tonegawa, who showed that joining DNA was a key mode of diversification for antibodies. In the mid-1970s, when Tonegawa's work was published—and when recombinant DNA methods were developed— I was seduced to become an immunologist. That brought me in contact with Lee because we were both trying to understand the mechanisms that generated antibody diversity. We were *really* trying to work out the molecular biology of a process that had much earlier been recognized by Sir Frank Macfarlane Burnet and Joshua Lederberg and enshrined as the clonal selection theory.

Lee attracted some remarkable trainees, notably Mark Davis. My equally remarkable trainees, including Fred Alt, and I found ourselves vying with the young hotshots under Lee's direction. Looking back, we and others gave life to an understanding of how immune cells can use DNA rearrangement to generate an immune system.

For me, understanding this process was a driving force that lasted years. For Lee, it was the beginning of a career that quickly moved into other realms. In particular, Lee saw that the huge variety of antibodies that allow us to be protected against viruses, bacteria, and fungi represented a myriad of protein structures. Early on, he recognized that automated equipment and methods were essential if we were ever going to unravel the complexity of the system. This sounds simple, but it was not a mainstream view at the time. The traditions of molecular biology had developed in an entirely different way. The field was dominated by small laboratories, by simple experiments often read out overnight, by dedicated investigators who combined complex theory with their revealing experiments to create deep understanding, usually of single biological mechanisms. The idea of automation to speed up a creative process was repugnant to senior molecular biologists. It turned out that they were seeing the past and Lee was seeing the future. He brought into his orbit engineers, machine builders, and many others who helped him create a new way to do science. In particular, he showed us how to find, wholesale, the structures of proteins and DNA sequences, providing tools that matched the complexity of living processes and opened new ways of doing biology.

But that was only the beginning. Lee moved from Caltech to the University of Washington in the early 1990s because his vision of the future was broader and more complex than Caltech—a boutique university—could handle. Even the UW, a huge state university with basic research and a medical school, was not able to satisfy his voracious appetite for science. In early 2000, he left to form an independent institute, the Institute for Systems Biology. There he has evolved a view of medicine and science moving to a new paradigm of maintaining health, enshrined as the four Ps of medicine: predictive, preventive, personalized, and participatory. He is involving people in their own health care. Wellness is his concern, not so much the treatment of disease.

Lee is a self-controlled person. Many find him hard to penetrate, hard to warm up to. But for me, Lee represents an important mode of doing science. Not only doing the obvious that we all do, not only doing the detailed investigation of systems that cry out for understanding, but constantly asking: "What do we need to get to the next level?" Sometimes he appears to be superficial, but he is always on to something, always striving to create the future. Often it is others who find their way to the next rung of the ladder of biological understanding, but they are hearing Lee and responding to his inspiring message. He is the pied piper. I believe that the advancements of science are created by the intersection of people who have different styles, different visions, different strengths. To this amalgam, Lee contributed a special set of capabilities, ones that are in short supply. Our science would be less powerful, less explanatory, if it were not for Lee's insights and his determination.

Lee's life story provides rich subject material for a biography. His experience reveals much to us about the character of those who drive progress, and science itself. Luke Timmerman has captured Lee's complexity and laid it out for the world to appreciate. Enjoy your encounter with this deep personality carried along by a fine biographer.

David Baltimore
President Emeritus
Robert Andrews Millikan Professor of Biology
California Institute of Technology

Foreword

Willing Shoulders:
A Personal View of Lee Hood

Only seven years after publishing his first paper, Lee Hood was a big deal in biology. He had already become synonymous with studying antibodies, using the protein chemistry that was common then. The unusual thing about Lee was that he wasn't afraid to embrace the engineering side of science. Bioengineering was barely a field, generally concerned with nutrition and plastic prosthetic limbs. "Genetic engineering" was a term hardly used, just before the dawn of recombinant DNA. In fact, "engineer" was not a term bestowed with glory or genius: it was around this time that rocket engineers renamed themselves "rocket scientists." Young biologists who wanted to develop better techniques in the lab were destined to become lowly technicians. Lee was already beginning to change that. He was making it socially acceptable (and, to some of us, irresistible) to engineer instruments for biology.

I first met Lee around 1978, when he visited the Harvard BioLabs and gave a talk in one of our small conference rooms. He was obsessed with DNA cloning and sequencing, even though all of his work was in proteins. Lee had just published a paper in *Nature* that did little to dispel the fact that he was an outsider in the field of DNA, while those of us he was talking to in the room were quite familiar—including the ultimate insider, Wally Gilbert. But somehow you could already see that Lee had a very attractive vision for where DNA could go: automation. Not everyone shared his vision, of course. As a grad student in Wally's lab, I had just written some software to read DNA sequences automatically using a drum scanner, and when I proudly showed it around, one person said, "Why do you want to automate reading? It's the only part of DNA sequencing that's fun."

But Lee already knew very well why, and he steadily began working to bring his vision to fruition. Lee had a simply stated set of four goals: to enable the automated reading and writing of proteins and of DNA. His greatest contribution to biology was aggregating teams to accomplish these goals. By July 1984, he published a paper in *Nature* titled "A Microchemical Facility for the Analysis and Synthesis of Genes and Proteins." It nailed three of the four quadrants. Two years later, the title of Lee's June 1986 *Nature* paper beckoned with the term "automated DNA sequence analysis." But, in his enthusiasm, he had oversold. I could not find any automation. The reactions and gels were done manually, the data came like chart recordings, and these were interpreted by pencil on the charts. Soon, however, collaborating with Applied Biosystems to commercialize a DNA sequencer, Lee nailed it. The quadrants were full. Yet the innovation continued: ink-jet printing of DNA arrays, next-generation sequencing, mass spectrometry for proteomics.

The enthusiasm Lee brings to biology spreads well beyond the engineering of instruments. He has helped establish more than a dozen biotech companies. He is happy to sit on companies' scientific advisory boards, some of which we work on together: Genomatica, Complete Genomics, Cellular Dynamics, MyMercuryRisk, Nabsys, and AbVitro. He is passionate about P4 medicine: predictive, preventive, personalized, and participatory. He has established a standard for sharing data worldwide. And yet he still hasn't forgotten about those coming up behind him.

Lee has for many decades helped create programs for teaching biology and biotech to high school students. He has written textbooks for undergraduates (which I used even if my courses did not assign them). When Lee first arrived in Seattle in 1992, he wanted high school students across the state of Washington to learn DNA sequencing hands-on. High school! Hands-on! How? Clearly not using the radioactive technique that was common then, nor using his own expensive invention, the ABI sequencer. Instead, he chose a nonradioactive version of my eight-year-old "genomic sequencing" method. I was stunned and thrilled. The most important thing to Lee was sharing his enthusiasm and encouraging people to participate. I would say that this still encapsulates Lee to this day.

George Church
Professor of Genetics
Harvard Medical School

Preface

Reporting for this book started in the fall of 2012, but it truly goes back more than a decade.

I first met Lee Hood when I was a kid reporter at the Seattle Times in 2001. He was in his early sixties, one of the world's best-known biologists and biotech entrepreneurs. He was building an institute doing some biology I didn't understand. His place in history was secure as the man who developed the machine that made the Human Genome Project possible, but he wouldn't rest on his laurels. He was a blur of energy, a force of personality stronger than any scientist or CEO I had interviewed, before or since. His entrepreneurial drive was the source of his greatness and his troubles: Bill Gates recruited him to the University of Washington, which he later quit in a huff. People either admired the man or despised him. I wanted to know more.

Over the next decade, I wrote dozens of stories about Hood. He was generous with his time. He made biology clear, understandable, and even adventurous—unlike the drudgery I recalled from high school. We had some disagreements over stories but nothing unusual.

Genomics, as I worked on this book, continued to advance at a breathtaking pace. Google-backed 23andMe raised public awareness of the genome's influence on health, cheap DNA sequencing technologies were improving at a rate faster than Moore's law, and a few promising drugs had hit the market. Drama also continued to follow Hood: his institute had financial troubles.

Hood had told a romanticized version of his story in public talks, but my goal was to dig deeper, to arrive at a nuanced, balanced portrayal. This book is not an officially authorized biography, but an independent work of journalism based on more than thirty hours of interviews with Hood, plus interviews with more than one hundred friends, family, and critics. I have reviewed thousands of pages of documents, including Hood's peer-reviewed publications, letters, memos, and internal e-mails. I traveled to his boyhood home in Montana, the place of his greatest accomplishments in Southern California, and his home in Seattle. While I begin the book chronologically, Hood's whirlwind career became too convoluted for that. The latter chapters are arranged by subject; I revisit certain time periods several times.

This is a biography of a scientist, not a history of science. Science and technology are practiced by human beings alive with hopes and fears like everyone else. My hope is that learning about this particular character helps us better understand what many have called the "biology century." "All history resolves itself very easily into the biography of a few stout and earnest persons," Ralph Waldo Emerson once said.

This is the story of one man's remarkable journey—one that has put a biological revolution in motion.

Chapter 1

Mr. Gates, I Quit

Two weeks before Christmas 1999, Lee Hood appeared to have it all. A loving family. Money. Fame. Power. He counted Bill Gates, one of the world's richest men, as a friend and supporter. Eight years earlier, Gates gave the University of Washington $12 million to lure the star biologist from Caltech in what the *Wall Street Journal* called a "major coup." Hood's assignment: build a first-of-its-kind research department at the intersection of biology, computer science, and medicine.

Even at sixty-one, the former quarterback of an undefeated high school football team could still do one hundred push-ups in a row. He ran at least three miles a day. He climbed mountains. He traveled the world to give scientific talks to rapt audiences. He slept just four to five hours a night. Apart from indulging in a luxurious art-filled mansion along Lake Washington, he cared little for the finer things in life, sporting a cheap plastic wristwatch and driving an old Toyota Camry. He still had the same wonder and enthusiasm for science he had as a boy.

Yet here, at the turn of the millennium, Hood was miserable. Just as his once-controversial vision for "big science" was becoming a reality through the Human Genome Project, he didn't feel like a winner. He felt suffocated. He had a new vision, a more far-sighted and expansive one that he insisted would revolutionize health care. He felt the university bureaucrats who were in charge couldn't see the vision and kept getting in his way. It was time for Hood to have a difficult conversation with his biggest supporter.

On a typically dark and gray December day in Seattle, Hood climbed into his dinged-up Camry and drove across the Highway 520 floating bridge over Lake Washington to meet Gates, the billionaire CEO of Microsoft. Hood came to say he had resigned his Gates-funded professorship at UW. He felt he had no choice but to start a new institute, free from university red tape. It was the only way to fulfill his dream for biology in the twenty-first century.

Gates was well aware of Hood's record of achievements and its catalytic potential for medicine. Hood led the team at Caltech that invented four research instrument prototypes in the 1980s, including the first automated DNA sequencer. The improved machines that followed made the Human Genome Project possible and transformed biology into more of a data-driven, quantitative science. Researchers no longer had to spend several years—an entire graduate student's career—just to determine the sequence of a single gene. With fast, automated sequencing tools, a new generation of biologists could think broadly about the billions of units of DNA. The sequences were obviously important, as they held the instructions to make proteins that do the work within cells. Thousands of labs around the world—at research institutions, diagnostic companies, and drugmakers—used the progeny of Hood's prototype instruments as everyday workhorses. George Rathmann, the founding CEO at Amgen, biotech's first big success story, once said Hood "accelerated the whole field of biotechnology by a number of years."

Hood had an unusually clear and far-reaching view for how biologists could fully exploit the new instruments. His enthusiasm inspired many bright scientists to devote themselves to his vision and to do their best work.

Now, at the start of the twenty-first century, Hood believed biology was ready for more ambitious goals. He believed traditional reductionism, looking at one gene at a time, was outdated "small science." The time was right, he argued, for scientists to look at hundreds or thousands of genes and proteins together in the complex symphony that makes up a whole human organism with trillions of cells. Biology, like physics, had an opportunity to become "big science"—fueled by big money, big teams, and big goals.

The way to tackle such complexity, Hood said, was through what he called "systems biology." It was a new twist on an old idea that involved bringing scientists together from biology, chemistry, physics, mathematics, and computer science. He would break every rule and custom of academia, if necessary, to make it happen. That was his way.

But it wasn't going to happen at the University of Washington. Abruptly, he quit.

On that gloomy day in December 1999, Hood wanted to break the news to Gates in person. Gates was curious to hear what was so important that it couldn't wait.

The men sat down in a couple of comfortable chairs in Gates's office in Building 8 at the Microsoft campus in Redmond. Hood got right to the point. He'd resigned his endowed Gates professorship at UW because the bureaucracy of a public institution would never be flexible enough to let him achieve his goals for multidisciplinary systems biology. Hardly stopping for breath, Hood barreled through his long list of grievances with administrators who didn't share his vision. He rhapsodized about the opportunity for systems biology.

The billionaire listened for a solid fifteen to twenty minutes. When Gates asked whether the dispute could be resolved some other way, Hood said he had tried for three years to set up such an institute within the university. Starting his own institute was the only way.

When Hood had said his piece, Gates cut to the heart of the matter.

"How are you going to fund this institute?" he asked.

"Well, that's part of the reason I'm here ..." Hood replied.

Gates interjected. "I never fund anything I think is going to fail." Hood was stunned.

He hadn't expected Gates to commit on the spot to bankrolling a new institute. But he didn't expect to be flatly dismissed. Gates was a logical thinker, not the impulsive type. He was a kindred spirit, an entrepreneur, a fellow impatient optimist. Years earlier, they bonded on a safari in East Africa; Hood listened to Gates talk about the "digital divide" as hippopotamuses grunted in the night. Often, Gates peppered his biologist friend with questions about the human immune system, widely considered to be the most sophisticated adaptive intelligence system in the universe. The recruitment of Hood helped raise the University of Washington to international prominence in genomics and biotechnology during the 1990s. Given that success, Hood thought he could talk his friend into providing as much as $200 million for a new institute.

Gates didn't give his institute a penny in its first five years.

On the drive home that day in December 1999, Hood wondered whether he had said something wrong, failed to make a case. But it was a fleeting emotion. Moments of self-doubt, to the extent he had them, were brief. He confided in his wife, Valerie Logan, the one person he knew would support him, no matter what. He brooded for a while. "It shocked and hurt me," Hood said. "It was a statement of skepticism from someone I had hoped would support me."

Others who were close to the situation understood why the meeting had gone badly. "Bill had not, at that time, been schooled in philanthropy," said Roger Perlmutter, a former student of Hood's who went on to run R&D at Amgen and Merck. "This gift to the University of Washington to create Molecular Biotechnology was surely the biggest thing he had done in philanthropy. It was all done to bring Lee here. And then in short order, it unravels? It was a kick in the teeth."

If Hood's first thought was that he had possibly damaged his relationship with his most important benefactor, his second thought was that his vision was right. He would find other support. He had some money already. He had shares in Amgen, the biotech company he advised from its early days, which went on to become one of the best performing stocks of the 1990s. He also made millions from royalties on DNA sequencers sold by another company he helped start—Applied Biosystems. Hood had other wealthy friends and companies he could call.

There was a lot to think about beyond science. Where to begin on starting a new institute? Even though he was hailed as one of biotech's great first-generation entrepreneurs, Hood had never played an executive role in running those enterprises. Now, he would have to act as a start-up CEO responsible for not just vision and fundraising, but day-to-day operations. He knew he wasn't a skilled administrator. He was impatient in meetings, lacked empathy, and made clear to all around him that he didn't want to hear any bad news. He had a habit of avoiding sensitive personnel matters, like whether to fire people. And those qualities had resulted in some spectacular failures.

None of that deterred him. Hood had always overachieved, going back to his childhood in Montana. He craved the adulation of others. He always found it on the football field, in the classroom, the lab, the lecture hall, and the boardroom. When people told him he couldn't do things, he would grit his teeth and prove them wrong.

Hood's son, Eran, once said of his father: "I always joke they should take the Tupac Shakur song, 'Me Against the World' and rewrite it as 'Lee Against the World.' They could take out the district attorneys and the crooked cops and put in university presidents and the medical school deans who just don't know, don't understand, his vision."

The path ahead was clear. Hood had to prove his vision was right. He would push himself around the clock, to the far ends of the earth, spend his last nickel. Just as he sought to prove himself as a boy, Hood would prove himself again. Nothing would get in his way.

Chapter 2

———

Growing Up
in Big Sky Country

The man who sparked the genomics revolution was born and raised in America's "Big Sky" country. Leroy Edward Hood entered the world on October 10, 1938, in Missoula and spent his boyhood moving among a handful of small towns in Montana. The son of an engineer for the local telephone company, Hood became the classic all-American boy of the 1950s.

At a tiny high school with just forty graduating seniors, students were encouraged to become well rounded, testing themselves in multiple activities. Young Hood took on everything with gusto. He wanted to be the best, and in this small pond, he usually was. His list of activities would impress even the most ambitious résumé-padding students preparing for college today. He participated in football, basketball, and track. He joined the debate club, school play, chorus, and Boys State, a student government activity that offered a chance to meet the governor. His teachers didn't see him as a genius. They saw a smart, earnest, hardworking boy with potential. They encouraged him to aim high. What little he knew about the world outside Montana was based on books and magazines from the local library. What he loved about Montana was the sense of wonder and self-confidence it instilled. "Montana is a wonderful place to grow up," Hood once told a local newspaper. "Its wide-open spaces give you a sense of freedom and a belief that you can do anything."

Missoula in the late 1930s was a speck on the map surrounded by mountains, forests, and trout-filled streams. Two natural treasures—Glacier and Yellowstone national parks—were a couple of hours away by car. Directly to the east lay the Great Plains with its wheat fields, scattered oil rigs, and cattle ranches. Seven Indian reservations were spread around the state, four of which were bigger than Rhode Island. The nearest big city was about five hundred miles away. When Lee Hood was born, Montana was one of the nation's largest states by area and one of the smallest by population. That hasn't changed.

When Hood's ancestors moved to Montana in the nineteenth century, they marveled at its beauty, much as the explorers Lewis and Clark did. Writers have long grasped for words to describe the place and the feelings it evokes.

"It is a land of the big sky," the Rev. George Harper wrote in 1952 to his family in Tennessee. "You feel little standing under it in the bright sunlight on a crisp cloudless morning, or lying on your back looking up at the night with stars so big and near, you feel you can reach up and touch them. You learn what humility means. But at the same time you feel big. There is no place for little thoughts, or narrow feelings. It's just too big a country for little things." John Steinbeck romanticized the place in his 1962 book *Travels with Charley*. "What Steinbeck saw in Montana was a blank canvas for the imagination of adventurous boys. From Lewis and Clark to

Teddy Roosevelt to Ted Turner, that's how America has always seen Montana," the author Brad Tyer said.

Although Montana didn't have the industry or universities of the East Coast, it nurtured curious minds and can-do spirits. Irving Weissman, a year younger than Hood, grew up in nearby Great Falls and became a famous stem cell biologist at Stanford University. The two scientists, who became lifelong friends, both believed the remote, landlocked upbringing was central to their success. No one in Montana thwarted their ambitions or told them they weren't good enough. "The one thing about people from Montana that I noticed was different from people in the East was that we had no idea of tradition or hierarchy," Weissman said.

Hood's birthplace of Missoula, featured today in outdoor magazines as an idyllic small city with mountains on all sides, was founded in the 1860s. A US Army fort was established in 1877 after white settlers asked for protection from Indian tribes. A little more than a decade later, the University of Montana was created. By the time Hood was born on a warm, dry autumn day in 1938, Missoula's population was barely fifteen thousand. The Great Depression was dragging on. Many unemployed young men, including Hood's father for a time, were hard at work in the Civilian Conservation Corps. They hacked through the wilderness to build roads and trails for recreation and economic growth.

The natural environment attracted adventurous souls, hermits, and jacks-of-all-trades. Montana is a place with extreme weather—scorching hot in summer, numbing cold in winter, with regular threats from thunderstorms and tornadoes. It was not an easy place to live. Settlers needed a certain resourcefulness and resilience. Hood's grandparents had both. They—along with Hood's parents—lived most of their lives in Montana. Leroy was named after his maternal grandfather, Leroy Clarence Wadsworth, who was known as Roy.

Roy Wadsworth, the son of a stonemason from Nebraska, came with his family to Montana in 1896 at the age of seven. Wadsworth spent the rest of his life in and around Red Lodge, a coal-mining town near a gateway to Yellowstone. Granddad Roy "was one of those guys who could do anything," Hood said. He was a stonemason and a carpenter. One of his proudest achievements was helping to build the Beartooth Highway, a picturesque stretch of winding mountain road that's now designated as a National Scenic Byway. When weather didn't permit invigorating physical labor outdoors, Wadsworth did a variety of odd jobs. At one time, he drove a chuck wagon around Yellowstone. Other times, he did less pleasurable work in the coal mines.

Granddad Roy was popular in the small-town social scene. He was tall at six foot five, and "cut a handsome figure" with neatly combed silver hair, according to a newspaper article about his retirement. "He had the ability to mix well with any crowd and many a 'gal' of that day could say with authority that Roy dances like a dream," the newspaper said. He had a playful, impish streak. "Roy is a very bright boy, but he watches me with one eye, and looks for mischief with the other," one teacher said.

Hood's grandfather loved to tell stories. He regaled his grandchildren with tales (most likely tall tales) in which he claimed to have met the mountain man Jeremiah "Raw Liver" Johnson. The mountain man, mythologized in a movie starring Robert Redford, was said to kill American Indians, cut out their livers, and eat them.

Roy Wadsworth married Florence Freeman in his hometown on August 9, 1914. It was a shotgun wedding. The couple's only child, Myrtle Evelyn Wadsworth, was born four months later.

Florence, as family lore has it, was a proper woman and a stick-in-the-mud. She wasn't overtly religious, but she lived by traditional mores. She once argued with her husband about whether to take the grandkids to see the circus, saying such entertainment was immoral. Florence was often grouchy or unpleasant. "Florence married Roy because she was pregnant," a granddaughter said. "It bothered her, and she took it out on other people." She was particularly unkind toward her daughter, Myrtle.

It didn't help when Myrtle became pregnant out of wedlock. She had to break the news that she was following her mother's pattern, getting married in a hurry. The news didn't go over well, and Myrtle eloped.

Myrtle had met Thomas Hood during a university theater production. She was twenty-three when they married; he was twenty-two. Two days after their son, Leroy, was born—a healthy seven pounds, four ounces—Myrtle mailed a happy letter to her parents:

> Sorry to be so slow, but things have been happening so fast. Leroy is really a grand baby. You should see us beam. I have only seen him once, but everyone says he looks like you. At least he has lots of black hair. Is he a darling. Oh! Boy. Wish you could see him. I am fine.

The newlyweds rented a house on the north side of Missoula. Their family came together quickly. A daughter, Doral, was born one year and nine days after Leroy. Another son, Myron, arrived two years later.

Thomas Hood was a talented electrical engineer and creative problem solver. He was also stubborn, emotionally aloof from his family, a heavy drinker, and occasionally mean-spirited. His children later came to believe their dad resented his family responsibilities because they prevented him from fulfilling his full career potential. He didn't provide much love or affection to his children. Granddad Roy Wadsworth provided it instead.

Hood's parents sent their three children off for long periods with the grandparents, who lived several hours away. Granddad Roy became an important figure in Leroy's life. And Leroy became his grandfather's trusty helper, working with him to construct buildings and roads during summers. "My granddad was, in many ways, my real father figure," Hood said. "He never had a son. He only had my mother. So he saw me as the long-lost son. He really loved all of us kids, and me especially."

It was for good reason that Thomas Hood struggled to show affection.

"Our dad had a very rough childhood," Doral said. "He was put downstairs in the basement with only bread and water, and locked in the dark basement for days."

The Hood side of the family had a complicated emotional history. Lee's paternal grandfather, Ned Hood, was born in 1889 in eastern Montana. There were significant troubles between white settlers and Native tribes. Ned grew up just a little more than a decade after the Battle of the Little Bighorn, also known as Custer's Land Stand. He lived almost his entire life within an hour or so of the site of that bloody battle.

Ned, an only child, did get an education. His father helped build a school and was one of five neighbors who "paid the teacher from their own pockets to insure the 3 R's for their children," according to Ned's 1959 obituary. Ned rode a pony five miles every day to school. He was the school's first eighth-grade graduate.

Ned married Clara Flanagan in August 1912, and their son, Thomas, was born three years later. But Thomas never really got to know his father. Thomas was a toddler when Ned left home to fight in World War I. There, injuries from a mustard gas attack cost him half a lung. When Ned returned after the war in 1918, he became an alcoholic and most likely suffered from post-traumatic stress disorder. Ned "was not a fit father because of all his war injuries," said his grandson Myron. After the war, with Thomas four or five years old, Ned and Clara divorced—an unusual thing in those days.

The family trauma set off a disturbing series of events for young Thomas. His mother married a man named Cecil Ryan and went on to have three more children. Cecil didn't like Thomas. He punished the boy by locking him in the dark basement, alone, for a day or two at a time—child abuse by any community standard. Eventually, Thomas was shipped out of the house. He was shuttled between the homes of aunts and uncles, not spending much time with his mother as a boy. For a while he was sent as far away as Missouri. Before graduating from high school, Thomas ran away from home. He worked odd jobs and ultimately put himself through college.

By all accounts, Thomas Hood was hard-driving, brilliant, arrogant, distant, and fiercely independent. Overcoming an abusive upbringing to become a successful engineer "was a spectacular achievement," Lee reflected later. Thomas was a stocky man of about five foot seven, whose thick-framed glasses and upright posture gave him the bearing of a man not to be trifled with. He also had a fun-loving, charismatic streak that he seldom displayed at home. With a few beers after work with his employees, he'd loosen up. "He could be the life of the party," Myron said.

Thomas studied physics at the University of Montana, where he met his wife. Myrtle got her bachelor's degree in home economics—a rare achievement for women then. Thomas studied hard during a golden era for quantum physics. But, a practical man focused on electrical engineering, he refused to take a required French course. He never graduated.

Lacking a bachelor's degree, Thomas took an entry-level job at the Mountain States Telephone Company. He spent his career climbing the corporate ladder, building communications infrastructure across the state for a company later acquired by AT&T.

The job required the family to make frequent moves. After getting started in Missoula, the family moved to the small towns of Ramsey and Butte, back to Missoula, and ultimately to Shelby for Hood's high school years. Each move was at least one hundred miles, which meant saying good-bye to friends and meeting new people. Once settled, Thomas spent long stretches working in the field. "My father was 98 percent devoted to work and 2 percent devoted to family," Hood said.

Lee's parents fought regularly. Tensions flared on weekends when Thomas was drinking, carousing, and staying out late. He could be reckless, sometimes driving drunk. He once narrowly escaped a head-on collision with his family in the car. Hood's mother, who didn't drive, urged her oldest son to learn quickly so that he could safely shepherd the family home down dimly lit, narrow roads at night. Long before getting his license, at age fourteen, Leroy Hood took responsibility as the designated driver for his family.

Hood's father was so legendary for his wild side that his reputation stretched across the state. On one visit to a mining outpost west of Missoula, Myrtle met a stranger. "Oh, you're married to Bulldog Hood," the man said. Myrtle had to ask what the name meant. Thomas earned it one drunken night when a sheriff's deputy tried to arrest him—and he bit the officer on the ankle. Sleeping off another binge, Thomas woke up one morning with a Mohawk haircut given to him by a mischievous cousin.

Hood didn't find his father's weekend exploits amusing. As the oldest son, he was very aware of the tensions between his parents. Many times, he served as a sounding board for his troubled, lonely mother. Myrtle told her son repeatedly that she might leave his father. "She confided in me all the time. All the time," Hood said years later. "She put way too much on me." Hood constantly worried he might have to drop out of high school to support his mother and younger siblings. "My mother threatened to leave him as long as

I can remember. All through middle school, all through high school. I never knew from one time to the next what she would do," Hood said.

Myrtle didn't leave. "I remember my mom saying all the time, 'Oh, I should leave him,' and my dad saying, 'If you leave, I won't support you in any way, shape, or form,'" said Hood's sister, Doral. "In those days, he could get away with that. She didn't have self-esteem, or she figured she couldn't make a living for herself. For the kids, she stayed."

Like many women of her generation, Hood's mother stayed home to raise her kids full-time. As much as she bent to her husband's will, she had strong desires for her own children. For one, she wanted them to be independent.

Growing up, the Hood children had plenty of free time to get out of the house, run, and play. As young as age six, they hiked and camped on their own. A favorite place to play was Mount Jumbo, an iconic 4,700-foot mountain near Missoula. They didn't have scheduled soccer games, intense violin practices, or the long list of other organized activities that today's parents coordinate in hopes of getting their children into Ivy League universities.

The Hood children were simply expected to achieve. Like her husband, Myrtle expected excellence. All of the kids strove to meet the high standards. By example, Thomas and Myrtle passed down their Montana-hewn values: Stand up to challenges. Don't quit when things get tough. Once you make up your mind to do something, do your best. Figure it out on your own.

The kids never had to worry about the essentials—food on the table, clothes on their backs, roof over their heads—but money was tight. They went to the library for books. Family vacations usually involved camping. They never went to exotic places like California. If the kids wanted something extra, they were on their own. "As a kid, I can remember going downtown and picking up beer bottles, because you could sell them for a penny, if you wanted to go to a movie," Doral said.

Doing well in school was a given. All of the Hood children were at or near the top of their classes. Formal education for the children began at the Prescott School, a two-story grade school in Missoula. Hood wasn't a child prodigy, but he showed solid ability. He wrote one note to his maternal grandparents in February 1946, at age seven:

> *Dear Grandmother and Granddaddy:*
> I have a new piece called "The Sea Spray." I have
> a catalog for a new electric train. We are having a
> Valentine box. I have sliding with the boys.
> Love and kisses and hugs.
> *From Leroy Hood*

At the bottom, he did a little math for them: $49 + 52 = 101$ and $67 - 32 = 35$.

Doral noticed her older brother separate from the pack at about age ten. "Up until fourth grade, I used to do just as well, or better, than Lee," she said.

Hood had typical boyhood fantasies about what he might do when he grew up. In middle school, Hood dreamed of becoming a pilot. One teacher discouraged him. "I remember him saying, 'It's kind of like driving a truck. Do you want to spend the rest of your life driving a truck?' For me, that was it," Hood said. "I certainly didn't want to be bored."

Hood's mother encouraged him to be ambitious. There was always more to do, more to learn. She encouraged her oldest son to play musical instruments. First came the piano, then the clarinet, and later the soprano recorder. Hood said he resisted at first, but "my mother was implacable."

Then Myrtle got pregnant again in her early thirties. In April 1947, when Lee was eight, Glen Thomas Hood was born. He had Down syndrome, a genetic condition that leads to developmental difficulties and mental retardation. Local schools and social services at the time offered little for children like Glennie. He was called a "mongoloid." Thomas was unable to fully accept the boy. Myrtle had a nervous breakdown.

From the time Glennie was a newborn until age two, he spent most of his days with his grandparents. As Myrtle recovered, the family prepared for a big decision. Should Glen live at home or be sent to a state institution? Hood, by then age ten, was old enough to know what was happening. "My mother was 100 percent for keeping him at home," Hood recalls. "My father was 100 percent for sending him away." As usual, his father prevailed. The family would periodically visit Glen, three to four hours away. Despite feelings of guilt and doubt, most of the family believed it was the right decision. Glen became an independent adult, able to hold a steady job. He died in March 2008, at age sixty, a few weeks after his mother.

Many years later, when Hood accepted the Kyoto Prize, the Japanese equivalent of the Nobel, he wondered about his brother's influence on his career choice. In his acceptance speech, Hood said he'd known in high school that he wanted to study biology, "but for reasons I never fully understood, perhaps in part due to my brother Glen, I was more attracted to human biology."

Once Glen's fate was decided, family life settled into routine again. Lee's father continued to bury himself in work. He may have partied hard on the weekends, but during the week, Thomas was a no-nonsense kind of guy. Talented and hardworking, he rose to a senior management position at the telephone company. Thomas had more than one offer to work at Bell Laboratories, the famed R&D facility in New Jersey. He was tempted, but his answer was always no. He had deep roots in Montana, and he wanted to raise his family there.

An engineer to the core, Thomas was inventive at home as well as at work. He once converted a coal-burning stove in the house into a sawdust burner, because he could get sawdust free from a local lumber mill. He rigged the home's wiring so that the radio would automatically turn off when the phone rang—and turn back on when the phone hung up. That technology became part of mainstream smartphone design sixty years later. Another time, Thomas took a small gasoline-powered engine from a washing machine and hooked it to a manual lawn mower. "He made the first power lawn mower anybody ever saw," Myron recalled. Thomas also loved to fiddle with clocks. He kept about twenty of them around the house. He would take the clocks apart, rebuild them, and synchronize them all precisely to international standard time.

Though he didn't engage his son in his tinkering, Thomas wanted Lee to follow in his footsteps, to be an engineer. One summer, Thomas took his son to Denver to take technical courses that his employees were taking. It wasn't an invitation. It was an order. Thomas wanted to push the teenager, and he wanted to show off his smart son to his employees. Lee gritted his teeth. "I really resented that. I didn't want to do it," Hood said. But he didn't dare defy his dad's orders.

His father was the undisputed head of the household and a strict disciplinarian. Lee insists he wasn't a victim of child abuse, as his father had been. But misbehavior brought out a board, or a belt, for spankings that would continue until Thomas knew the child was feeling true pain. "As I got older, I became more and more determined not to cry. It was our battle of wills," Hood said. If Thomas meant those spankings to instill fear, he succeeded. "I think the whole family was kind of afraid of my father," Hood said. Yet Thomas was less harsh with Doral. Lee resented his sister's favored status. One time, when Lee was in elementary school, his initials were scratched into the paint of a company-owned truck. Thomas "was really furious, and he really came down and spanked me," Hood recalls. "I told him I didn't do it. I told him it was my sister. For whatever reason, not much ever happened to her." Myron, who was more passive than Lee, also tended to clash less with his father.

Thomas and Lee had perhaps too much in common. Both were strong-willed personalities, never the type to back down from a fight. They were "almost identical" in their brilliance, creativity, work ethic, and stubbornness, Lee's siblings said. "Lee got a lot of the drive from my dad," Doral said. "Maybe [part of the conflict] was that Lee threatened him, because he could do so much, too. But my dad never discussed emotions or feelings, or anything. He was a very stoic person."

Thomas didn't talk about his traumatic childhood, but as a father, he made sure his kids could stand up to harsh experiences. Myron said his father's "tough love" parenting style made him think of the Johnny Cash song, "A Boy Named Sue."

Lee Hood's most vivid memories of his father were that his dad didn't attend more than a couple of his high school football games, didn't drive him off to college, and didn't attend his graduations. He never gave a word of advice. Not once did he tell his son he'd done a good job. "He was never really interested in me as a person, even later when I was successful," Lee said.

The distance, and influence, of the father is evident in the tone of a letter Hood wrote when his dad retired. It reads:

> I suppose it is appropriate at a retirement dinner to
> think about the past and how it has influenced the
> future. I do have some happy memories of how you
> and your association with the telephone company

have influenced my life. First, in retrospect, the
frequent moves that we made were, I think, good
for all of your kids. We had to meet new people
and situations and learn to adapt. In today's highly
mobile society, that was an important opportunity.

Second, my opportunity as a high school student
to take a telephone company course for a second-
class radio operator's license in Denver was a
unique experience. I got a much broader exposure
to a certain aspect of science than I had before. It
was a heady experience being treated as more or
less an equal by the grown-ups. It certainly was an
opportunity to spend an extended period of time
with you.

Finally, I remember the summer job that you got
me in Helena as an installer. There I have fond
memories of learning some about the many things
that I knew little about, but more important, I
enjoyed the people I worked with. I guess one overall
impression I have about your friends and associates
in the telephone company is that they were fine
people. You must feel very fortunate to have had
these kinds of friends.

In closing, I hope the future will be as rewarding and
exciting as the past has been for you. In my mind,
there is no doubt that you'll be, if anything, more
productive and busy in retirement. You have our
best wishes for this happy occasion and our love and
support for the future.

It's hard to tell whether Hood truly enjoyed the "opportunity to spend an extended period of time" with his father. Ironically, Thomas was proud of Lee and even lived vicariously through him. He read newspaper clippings about his son's achievements. "Everything Lee did, whether it was playing football or anything else, my dad would brag about," Myron said. "But he would never brag to Lee."

Lee pushed himself as a child to do just one thing that would win his father's affection, or at least earn a positive comment. As an elite student, as a Caltech professor, as a visionary scientist—he sought his father's approval. He didn't get it. "Our dad wasn't capable of giving affection. He never was," Myron said.

But others began to recognize Lee's achievements. In his high school years, Lee started to get the affirmation he sought. The community of Shelby—its students, teachers, coaches, fans of its championship football team—fully embraced the town's new golden boy. Once Lee had a taste of affirmation, he couldn't get enough.

Chapter 3

The Golden Boy of Shelby

Of the four Montana towns where Hood lived as a boy, Shelby was most important. The community put the teen on a pedestal.

The family moved there when Hood was about to enter eighth grade. This was not another picturesque mountain village. It was a rail hub, population 3,058, about thirty miles south of the Canadian border and eighty miles east of Glacier National Park. The wind howled down the Rocky Mountain foothills, strong and steady enough to attract massive wind turbines in years ahead. A few ranches, feed mills, and grasslands were scattered on the periphery. Culture in this rural place consisted of rodeos and demolition derbies. Shelby's lone moment in the national spotlight, thirty years earlier, was a dubious one. Thrilled by the discovery of oil in 1922, some ambitious local bankers and oilmen sought to put Shelby on the map. They lured boxing champion Jack Dempsey, a national hero, to defend his heavyweight title at a thirteen-thousand-seat wood arena they built just for the event. The fight was a fiasco, as not enough paying customers showed up. Shelby's oil reserves were limited, and several banks that had helped finance the fight soon went bust. The town's dreams of prosperity went unfulfilled. Today, near the site of the original fight, stands a Pizza Hut, a truck stop, a small casino, and a billboard spray-painted with a skull and crossbones: "Not Even Once. Meth Kills." Many downtown storefronts haven't been updated in decades. A veterans group found it tough to raise a few thousand dollars to maintain a giant American flag installed on the west side of town.

But when Hood lived in Shelby, during the Eisenhower years, the town enjoyed a period of postwar prosperity. Decades later, Hood was nostalgic about the place. He attended several class reunions, including his twentieth in 1976. "Shelby was a great place to grow up," Hood wrote in a speech outline. "Small school. Exceptional teachers. Exceptional freedom. Lack of pressure. Confidence to do things. Enjoy life. Human values."

Hood was senior class president, editor of the yearbook, and valedictorian. He didn't ace every class, but his record was nearly spotless, with only a couple of Bs and one C in chorus. Back then, the local newspaper published the names and grade point averages of every student on the honor roll. Hood and one other boy were the only ones with straight As the spring of their senior year. Classmates voted him "best all-around."

Hood was one of the youngest members of his class, still five months shy of turning eighteen when he graduated in May 1956. Even though he was physically small, he played football all four years. These became fodder for sometimes-exaggerated stories about his football glory days—stories Hood told throughout his scientific career as a way of relating to nonscientists. Hood was a starter on defense for two years, then started at quarterback as a senior. Not only was Shelby undefeated during Hood's season at the helm, but it dominated opponents by an average score of 40-6. Still, it was a talented team, not a one-man band. "I was no Johnny Unitas," Hood later told a Montana newspaper. "I was good, but not great, and we had several kids who were terrifically speedy." Home games were social events. As quarterback, Hood was naturally the center of attention. Nearly sixty years later, classmate Arlene Littlejohn still remembered Hood's ability to throw the ball and make quick decisions. "He was so smart, he could read the defense," she said.

In the classroom, Hood was guided by a handful of talented, caring teachers. One of them changed his life. Cliff Olson, a science teacher and the school principal, was an educated man in a remote place. Olson got a bachelor's degree from the University of Montana, then a master's in meteorology from the California Institute of Technology in 1947. It was a stroke of great luck that Hood crossed paths with such a man. Olson urged the boy to apply to Caltech, one of the elite centers of science and technology in the world. Hood's father preferred he stay close to home and attend Montana State University, which the family could afford. But Olson was insistent. When Hood considered Carleton College in Minnesota—a well-regarded liberal arts school—Olson intervened. He came over to visit Hood's parents at home, personally making the case for Caltech.

Caltech was famous then, as now, as a hotbed of science and engineering. It was home to legends like Linus Pauling and Richard Feynman, and to famous visiting scientists like Albert Einstein. Olson vowed that, if he ever had a chance, he'd steer his brightest students there. At least five students from Shelby went to Caltech over a ten-year period when no other students from Montana got in. "That's really a striking demonstration of what good teachers can do," Hood later told a science historian at Caltech. He realized his good fortune. "Serendipity is an unbelievable force in what happens to us," Hood said. "You have to take advantage of it."

Years later, Hood marveled at the relationship he had with his teachers. "What was true of all of them is that they treated me as an equal," he said. "We'd go out and have a Coke, and talk about what was important in life. I had a feeling I was talking to a peer and not a teacher, in many ways. They all pushed this idea that you should be ambitious in your goals."

Olson in particular wanted to challenge his smart students and keep them from getting bored. He once asked Hood, as a senior, to step up in front of a sophomore biology class and teach. It was one way to push the boy out of his comfort zone. He needed something. Shelby didn't even offer calculus, meaning its college-bound students were at a disadvantage to those from bigger public and private schools. Never one to shrink from a challenge, Hood went to the library to get copies of *Scientific American*, including one that discussed the discovery of the DNA double helix. Learning biology, at least well enough to teach sophomores, stuck. Partly, it forced Hood to learn the material himself. It also forced him to become comfortable conveying his thoughts clearly in front of an audience. Both skills became critical to his success.

But Hood didn't become fixated on a career in biology through the experience. He didn't have some epiphany after a magical high school experiment. Hood was a high school freshman in 1953 when Watson and Crick discovered the DNA double helix. Word of a revolution in molecular biology was only just beginning to spread. *Scientific American* ran stories of the electrifying discovery and the many potential pathways for investigation, and Hood read them with curiosity. It didn't make a huge impression on him then. His transcript lists "medicine" as his vocational interest. In the yearbook, he wrote that his ambition was to be a psychiatrist. Brimming with energy, he didn't quite know how to channel it. His senior yearbook quote reveals a budding ego: "I'm on the brink of a great career—somebody push me off."

Hood was a brain, a jock, and a bit of a square. That's how classmates Arlene Littlejohn and Dwayne Irvin recall him, fifty years later. Like many boys, Hood had a short buzz haircut. He grew to stand five foot ten, with a lean, athletic build. Photos from those days show his deep-set green eyes, and the trademark piercing gaze that photographers captured repeatedly throughout his career. Even though underage drinking was a primary source of mischief, neither of Hood's classmates remembers him joining the keg parties down by the Marias River. When the cops came, the kids would run away through the fields.

Local teens might bend or break the rules, but for the most part the 1950s were a conformist period of sock-hop dances and conservative "ducktails and bobby socks" dress. Teachers, parents, coaches, and other authority figures commanded respect. Despite his record of clashes with university administrators as an adult, Hood was an obedient son during his teen years. He didn't drink or do drugs; he played by the rules. You could say he was a Boy Scout, except that officially he never joined.

Irvin delighted in one story about his oh-so-responsible classmate. Just weeks before graduation, the senior class went on a chaperoned "Skip Day," designed to let students relax at beautiful Flathead Lake. A couple of boys broke into a railcar and stole a dozen cases of beer to share. Hood didn't join the party. That same trip, another student got a little too aggressive when pulling a cord to start an outboard boat motor. The motor flipped into the lake and sank about ten feet to the bottom. It wasn't Hood's fault, but he sought to make it right.

"I remember Leroy standing there in his shorts on the dock, saying the water's clear, and it wasn't too deep. He thought he'd jump in the cold water and get it. He stepped right up to the plate to help out. That's the way he was," Irvin said. (Another classmate fished the motor out.)

Hood could be spontaneous and fun. He took dance lessons with Littlejohn—the samba and the rumba—and did well enough to perform for community organizations. "One night, we had a dance performance," Littlejohn said. "There was a prom in Sunburst, and Leroy says 'Hey, we're all dressed up, why don't we go to Sunburst and go to their prom?' So we crashed their prom."

One night of their senior year, Hood and his friend Stanley Underdal sat outside on the hood of a parked car and mused to their dates about the future. They looked up at the stars and told the girls that America would someday put a man on the moon. The girls howled with laughter at the fanciful notion. Five years later, President John F. Kennedy set that goal.

Some of Hood's most important education came during summers spent with his grandfather in the Beartooth Mountains. Granddad Roy lived near the Continental Divide, in one of North America's more geologically interesting landscapes. There, Granddad operated a small camp resort. For decades, Ivy League geologists came to the retreat in summers. Roy Wadsworth sought to make his visitors comfortable, building a road and cabins with basic amenities. Hood helped around camp. He reveled in the company of the bright students who accompanied the professors. Somewhat awestruck, Hood began to think more seriously about a life in science. He also got a vicarious glimpse of life outside Montana. "I got to talk to kids from Columbia, Princeton, and Yale," Hood recalled. "It was a different world."

Driven in part by the desire to test himself against tough competition, and encouraged by his grandfather and the visiting professors, Hood started thinking bigger. The summer of 1955, after his junior year, he entered the Westinghouse Science Talent Search. The Westinghouse, now known as the Regeneron Science Talent Search, has long been one of the most prestigious awards in the United States for high school students. Dating to 1942, the contest is judged by leading scientists. This honor often has served as a precursor for the Nobel Prize and the Fields Medal, the highest honor in mathematics. Taking a cue from the visiting geologists, Hood submitted research on a nearby rock formation of interest to oil prospectors.

More than 20,800 students applied for the Westinghouse that year. Hood was among the forty finalists—rare for a student from Montana. He won an expenses-paid trip to Washington, DC, where he listened to talks by scientists, took field trips, and presented his work to the judges.

The community of Shelby was so proud, the high school band played and five hundred people—one-fourth of the town's population—showed up for a send-off. Memories of that adulation stayed with Hood his entire life.

When he got on the Empire Builder train for the three-day trip, it was the first time the seventeen-year-old high school senior had traveled any significant distance. He went alone. When Hood arrived in the nation's capital, he was again reminded that he wasn't the smartest kid in the room. Records from the 1956 Westinghouse competition don't say exactly where Hood finished, but he wasn't in the top three, or the next ten. Hood recalls:

> The train ride was one of the exciting adventures
> of my life. You've never been away, and going back
> there all by myself. When I met the fellow winners,
> I was a bit intimidated by how smart and broadly
> knowledgeable a lot of these people were. ... After
> that, I knew I was good but wasn't in the upper-
> upper echelon of really smart people. That's an
> interesting topic, because at Caltech there were
> really smart people who never really accomplished
> that much in later life. It's interesting about what it
> takes to be successful.

Corlie Dunster, a social studies and history teacher, made this point to Hood, trying to prepare him for the wider world. She wanted him to know he'd be successful because of his enthusiasm and work ethic. "She said, 'You know Leroy'—I was Leroy then, rather than Lee—'I've had two or three students that are smarter than you, and you should remember that, when you go away to Caltech, because you're going to find Caltech much harder than you ever suspect,'" Hood said in an oral history. "And so I was kind of—I mean, I was hurt by this, and I didn't understand it. But I think in retrospect, I do understand some of the message she tried to convey."

Later in life, as Hood met people from around the world, he appreciated how lucky he was to have been nurtured by a number of caring adults in a small community. "I can just see myself being utterly lost in one of these big suburban high schools," Hood said in his Caltech oral history. "Maybe in the really big places, like New York City, super good people can go through Bronx High School of Science and end up being super. But it's small towns, like the one I was in, that can take people of more average talents like myself and give them the confidence that they can use their talents effectively."

As a teen, Hood certainly didn't see himself as having "average" talents. But being around the Westinghouse winners and the Ivy League geology students drove home a clear lesson. If he was going to play at that intellectual level, he would have to work extremely hard. "Success is a function of effort," one teacher reminded him in his senior yearbook.

Hood had work ethic in spades. His brother Myron recalls being amazed by his study habits. The boys shared a basement bedroom. "Most every night when I went to bed, his light would still be on, reading," Myron said. "Every morning when I woke up, his light would be on again. He was reading." When other families were getting television sets and watching popular early TV shows like *I Love Lucy*, Hood and his siblings were reading. Even though they could afford one, the family never had a television the entire time Hood lived at home. (Myron was relieved when his brother graduated and the family moved to Helena, so he could finish high school in a place where he wasn't known as "Leroy Hood's younger brother." He would go on to become a mathematics professor.)

When asked decades later what his most important courses were in high school, Hood didn't mention math, physics, or biology. He pointed to typing, which enabled him to efficiently communicate his ideas in writing, and debate, which helped him verbally convey scientific ideas with clarity and style.

> Debate for me was really transformational. When
> younger I was a little bit shy and hesitant. I had
> a terrific debate instructor. He set how I thought
> about speeches the rest of my life. He said the most
> important thing in giving a talk is audience contact.
> You learn what you want to say, you don't use notes,
> and you look them right in the eye. It's really true. If
> you do that, they have the feeling you're there with
> them, as opposed to you just reading off slides or
> your notes.

Hood made eye contact with his future wife, Valerie Logan, at a theater competition for Montana high school students. It was held in Missoula their junior year. Hood had played the part of a henpecked husband in a Shelby school play and won an award for his acting. Valerie joined the competition on behalf of her school in Cut Bank. After the plays, there was a mixer. Hood asked Valerie to dance.

The two then went their separate ways, but Hood was transfixed by the spirited girl. After the dance, they were reintroduced by a mutual friend. Valerie flashed her independent spirit one night after Hood led Shelby's football team to a 59-12 victory over Cut Bank. "That was the only date she broke with me the entire time we dated," Hood said. "She didn't show up. She said, 'You ran the score up needlessly.'" Valerie would keep Hood in check for the rest of their fifty-plus-year relationship.

By the spring of 1956, Hood was ready for college entrance exams. Caltech's requirements were daunting. An advanced math course was expected. Hood hadn't taken calculus, but he had studied advanced algebra, trigonometry, and solid geometry. "You'll do fine," Olson assured him. "Don't worry about it." Hood said later, "I suspect I got one of the lowest scores of anyone who ever got into Caltech, at least in advanced math."

Getting accepted was a thrill. The "great career" Hood dreamed of was now within reach. Hood's father's concerns about the cost of Caltech were allayed by a scholarship from the General Motors Foundation. The scholarship, which would pay most of his tuition, room, and board, was renewable for all four years.

Thomas Hood gave his son the family's second car, a 1950 Ford sedan, to make the twelve-hundred-mile trip to Pasadena. The two-day journey across long, desolate stretches of Nevada desert gave the boy plenty of time to think. It was a long time for a teenager on the road, across some of the most deserted stretches of America, with few hotels, gas stations, or consistent radio reception. He would have enjoyed some company. He wished for some moral support from his father as he left for the big, wide scary world. But he was on his own. "My father said he didn't have time to take me," Hood recalls.

Driving across the mountains, basins, and valleys of the West, Hood left the sheltered environment of Shelby, Montana, for a much more diverse, urban, pressurized place. He knew he didn't want to be an engineer like his father. He wanted to chart his own course. He wanted to do something great. He knew he was smart, but not the smartest kid in the world. If he was going to make it, he would have to drive himself harder than ever.

Chapter 4

Welcome to the Big Time

Arriving in Southern California, Hood immediately felt the sting. His eyes burned. It was the smog: a combination of smoke, fog, and airborne pollutants. That year, 1956, a survey found that nine out of every ten local physicians said they recognized the symptoms of "smog complex"—irritated eyes, cough, nausea, and headache. The air was so foul in fast-growing Los Angeles County that Hood would get coughing spells after football practices. "It was about three weeks before I saw the San Gabriel Mountains," he said. Those mountains, just a few miles away, dominate the skyline on clear days.

Just before turning eighteen, Hood was in unfamiliar territory. California was socially conservative, voting to reelect President Dwight D. Eisenhower by a landslide. But it was nothing like his boyhood surroundings. The Golden State was booming. Millions of new people, diverse people, were moving to its warm and sunny cities.

Quite a few were seeking some combination of fame, fortune, or freedom to pursue their individual ideas of happiness. The kid from the sticks had culture shock in Pasadena. "I don't think I had ever seen a black person. I don't think I had known someone Jewish. We were incredibly sheltered," Hood said. "There were lots of Jewish people at Caltech. I remember being a bit surprised they didn't look different."

The California Institute of Technology was founded in 1891 as a small technical school. It took less than twenty years to rise to prominence, under the leadership of astronomer George Ellery Hale. When Hood got there in 1956, the institute was still an all-male school. It had a reputation as an exceptionally competitive and demanding academic environment. This was a place where great men did great things. Like its rival MIT to the east, Caltech aimed to be the best in science and technology. "Devotion to the goal of being at the forefront both of science—which means understanding nature—and of technology—controlling nature—has characterized Caltech from its beginning," former president Harold Brown once wrote. Another former president, Marvin "Murph" Goldberger, said the institute's success was based on "absolute unwillingness to compromise on excellence." He added: "To be accepted as a student, you need near-perfect SAT scores, so you are immediately among the elite. Invitations to join the faculty come only after the most excruciating examination. There are no good [faculty] appointments at Caltech—only superb ones."

It was no exaggeration. Physics was king under the leadership of Robert A. Millikan. He arrived in 1921 from the University of Chicago, where he was known for his famous "oil-drop" experiment that was the first to precisely measure the charge carried by electrons. His work successfully established the atomic structure of the fundamental unit of electricity. Two years after arriving at Caltech, Millikan won the Nobel Prize in Physics for those accomplishments and for proving the existence of the photoelectric effect—a phenomenon, predicted years earlier by Albert Einstein, in which metals emit electrons when light shines upon them. Millikan was a charismatic public figure, too. He became, as one writer later said, "a founder, first leader, first Nobel Laureate, and all-around patron saint" of Caltech. Bright minds and warm weather helped Caltech attract a rotating cast of star visiting scientists, including Werner Heisenberg and Einstein. The institute continued to draw giants in their fields. In physics, that included Richard Feynman and Murray Gell-Mann, who were at the top of their game in the 1950s. Linus Pauling was the dominant force in chemistry. Biology, an emerging science in the early twentieth century, was established as a Caltech department in 1928. Caltech gained instant credibility in the field by recruiting Thomas Hunt Morgan, a fruit-fly geneticist from Columbia University. When Hood got there in 1956, the biology faculty was loaded with stars. Some of them were George Beadle (who discovered that genes act by producing enzymes), Max Delbrück (an influential figure to the first generation of molecular biologists), Ray Owen (who discovered immunological "tolerance" in which the immune system distinguishes between friend and foe), and Edward Lewis (a founder of developmental genetics who also studied radiation's role in spurring gene mutations and cancer).

Caltech was then, and remains today, a small place. Passing motorists can easily miss the 124-acre campus, roughly the size of a typical residential subdivision in Southern California. There were only 180 freshmen in Hood's class. They were warned in orientation that life was about to get hard. More than half of the incoming freshmen were valedictorians. Here, students were graded on a bell curve. The math was simple: many of these young men were about to become average for the first time.

Like all freshmen, Hood had a lot to juggle. He had to settle on which courses to take, which extracurricular activities to join, whom to become friends with, and how to earn spending money. With relatively little free time, Hood had to develop even more disciplined study habits. But first, he had to be chosen by one of the four student residential houses on campus.

Caltech didn't have fraternities, but it did have an organized social structure. Students got their first taste of it at an orientation camp in the San Bernardino Mountains. They played sports, listened to faculty lectures, and had late-night bull sessions. Most importantly, freshmen went through a "rotation" process to determine where they would live. It wasn't quite like pledging to a fraternity and enduring the sometimes-cruel hazing rituals. But the newbies were sized up in a rather unnerving process by upperclassmen from each of the four houses. The older students were trying to determine whether the new recruit would fit in their house, where he was likely to stay all four years. The freshmen had some say, and they were on the lookout for compatible roommates. Hood was assigned to Blacker House, a place with a reputation for housing free spirits.

That fall, the Caltech student newspaper described the mating dance:

> There is much about rotation that, however
> necessary, is not emotionally appealing to many
> people. There is a great deal of superficial impressing
> and being impressed. This may be extremely
> difficult for some more sensitive individuals who
> would appreciate a low-pressure routine. Rotation
> combined with the rest of the new environment here
> at Tech could make life a trifle trying at times. This
> is probably the first time many have been roommates
> with some guy they didn't know, been served dinner
> by their schoolmates, or worked a Strong problem.

> We can only offer our sympathy to these freshmen
> and explain that there is a job to be done.

Hood hadn't worked a Strong problem (an unsolved mystery in particle physics). He was weaker in technical subjects, especially math, than some of his Caltech classmates. Many freshmen were able to skip first-year math and a few science courses. Not Hood. "People like me from backwoods Montana were at an enormous disadvantage," he recalled. One classmate that fall tested Hood, asking whether he knew about tensor analysis, a mathematical framework used in physics to understand concepts such as Einstein's theory of general relativity. Hood admitted he was out of his depth. "Boy, you're really going to have trouble around here," the other student said. Professors were intimidating, too, often more interested in their research than in teaching. The first couple months, all-night study sessions were the norm. It was, Hood said, a "depressing experience."

Students were terrified of failure. Mark M. Davis, a Caltech graduate student in Hood's lab in the 1970s, recalled the culture. (Davis later became a Stanford University immunologist and Howard Hughes Medical Institute investigator.) One Caltech boarding house, he said, had a mural that depicted a pilot and copilot looking out their windshield, staring straight ahead, and seeing nothing but the Millikan Library—an intimidating concrete monolith and the tallest building on campus. We'll never escape, the mural seemed to say. Davis described the Caltech environment as callous and isolating:

> At most colleges, the reason most people treasure
> their experience is they grew up a lot. You're on
> your own, away from home, navigating life, having
> fun. ... I went to Johns Hopkins, and it has a lot

of premed [students], who have some pathologies
that go with that. But, at the core, the attitude is
"we respect excellence" and you can be excellent at
anything. That's the goal. At Caltech, the culture is
very different. If you were a theoretical physicist,
if you were going to be the next Richard Feynman,
then the world was your oyster and they love you.
If you fall short of that, they really don't care very
much. ...

Here [the freshman] were: their parents were
proud, they were going to this great place—and they
were mean to each other. In some places, in some
circumstances, maybe they'd band together and
come closer and support each other. I don't think
that was happening in that culture for the most
part. People had friends, but they got damaged. Not
everybody, but some sure did.

Hood responded to the daunting course load by burying himself
in his studies. The gifted young classmate who had tried to intimi-
date Hood over tensor analysis, on the other hand, flunked out. He
spent too much time playing cards. "There were a few students who
were just orders of magnitude smarter than everyone else, and I
wasn't one of those students," Hood said. "Some of those geniuses,
in the end, didn't end up accomplishing as much as the rest of us
who worked really hard and were enthusiastic and engaged."

Whatever his disadvantages, Hood had advantages, too. Being a
debater, actor, and football quarterback gave him an outgoing personality and a reserve of self-confidence. Social awkwardness among
classmates was so common that Caltech had a formal program to
help freshmen meet "local belles" at nearby colleges, according to

the student newspaper. But Hood was comfortable as the center of attention. He was soon elected president of the freshman class. "I think I was more socially skilled than about 80 percent of my colleagues," he said. His infectious enthusiasm spread. By now, Hood no longer went by his full first name of Leroy, which his mother and grandmother preferred. At college, he was Lee. "The class of '60, under the leadership of President Lee Hood, has made a name for itself in many ways. Due to its unity, enthusiasm, and ability, it has been one of the strongest freshman classes that Tech has yet had," the yearbook said.

Hood found a friend to commiserate with about the academic pace, the workload, and the indifference of professors. Eric Adelberger became his best friend and roommate all four years. Adelberger—who went to a rigorous public high school in Arlington, Virginia—seemed to have little in common with the quarterback from the rugged West. The roommate was shy, didn't care for sports, and thought of himself as an "outsider." What they had in common was self-confidence, an intense work ethic, and a love of nature. When they needed a break, they'd drive up Highway 395 into the Sierra Nevada mountains. Hood and his friend, roped together, began learning technical aspects of rock and mountain climbing. Hood radiated enthusiasm for the outdoors. He and his friend talked about the places they wanted to go, things they were curious about in life, and what they thought really mattered. In the wide-open outdoors with his friend, Hood confided his complex feelings about his father. Throughout his career, Hood sought spiritual renewal in the fresh air of the mountains. "He was willing to try things and be open to various adventures in life, and talk openly about things," Adelberger said. "He really had a thoughtful side."

Adelberger went on to a distinguished career in physics at the University of Washington. He was elected by peers into the prestigious National Academy of Sciences. Talented and driven as he was, Adelberger said he paled by comparison with his roommate. He recalls:

> Lee had an absolutely tremendous drive. I remember when he was playing football at Caltech. He'd work himself into such a state where he'd hate the other guys. It was foreign to me. Very intense. I thought it was a game. He thought it was life and death.

Tenacity helped Hood forge through freshman year. He got Bs the first trimester, except for an A in chemistry. Second trimester, his grades improved. By the third term, Hood was finding a groove. He got As, except for a B in physics. Students were required to take two years of courses in math, physics, and chemistry as part of a famously demanding core curriculum.

More than fifty years later, long after administrators tried to curb competitive excesses by switching to a pass/fail system for freshmen, Caltech is still considered a unique crucible for undergraduates. "Typical students work and attend class for about ten to fourteen hours a day, leaving little time for other activities," says the 2006 *Insider's Guide to Colleges*. "Not surprisingly, sleep is usually the first thing to go. 'It's not unusual to see students up until 4:00 a.m., even frosh taking pass/fails—sleep is a very precious commodity,' said one student." The school's elite reputation continues. Three straight years in the early 2010s, Caltech was ranked No. 1 in a global ranking of two hundred universities, based on data from Thomson Reuters.

Hood was gifted with unusual powers of concentration and stamina. During final exams, he would go through the cafeteria in the morning, load up on sandwiches that he could take to the library, and disappear. Having a daylong food supply at his desk kept his energy up without wasting time on formal meals. "He worked much harder than I did," Adelberger said. "I started eventually adopting some of his practices. It did wonders for my GPA."

The drilling in hard math, physics, and chemistry—as well as diversions into well-rounded humanities courses—served him well. While many young biologists feared math and the physical sciences, and so avoided those courses, there was no escape for Hood. Later in life, he was thankful for the grounding in tough quantitative sciences. The training enabled him to form a broad worldview outside his discipline. And it gave him the confidence to compete in the complex new world of molecular biology. "He really had a leg up on molecular biology because of his Caltech training," said Daniel Kevles, a science historian at Yale University who became friends with Hood in the 1980s when they served together on the Caltech faculty.

Caltech wasn't all work. Hood ran track and sang in the Glee Club. He gave guest lectures to high school students at the YMCA—the start of a lifelong interest in sharing the wonders of science. And, of course, he played football. It was his most cherished extracurricular activity; he played all four years. But it was nothing like his high school glory days. Instead of playing quarterback, Hood played running back on offense and safety on defense. The positions require a player to withstand many violent collisions each game. At five foot ten and about 175 pounds, Hood might have been too small to play in these positions in big-time college football, but this wasn't big-time college football. Caltech sometimes struggled to fill a roster. It played home games in front of sparse crowds at the cavernous Rose Bowl. Even against weak competition, Caltech was a loser. Hood was disappointed. "Our high school team could have easily beaten our Caltech team," he said.

As social chairman of his house during his sophomore year, Hood arranged dances and other events with girls from nearby private colleges. This was no small duty, given that female undergraduates weren't admitted to Caltech until 1970. Hood had his opportunities with girls. But nothing lasted long. He stayed in touch with Valerie Logan, his high school sweetheart more than a thousand miles away.

Hood recalls going to music shows at the Hollywood Bowl, hitting the beach, and seeing movies with friends. Caltech students also had a long tradition of pulling off elaborate pranks. Once, Hood and a few friends plotted a mock bank robbery. They cased the joint, dressed in topcoats and hats. Hood dropped a map of the bank on the floor on his way out. Three days later, as the pranksters showed up to pull off their mock robbery, they were escorted out by security. Later, on campus, an FBI agent gave them a scolding. Real robbers who matched their description had been hitting local banks. They could have been in danger, the agent said. That only made the prank more exciting. "It was a lot of fun," Adelberger said. "Fortunately, we were at Caltech, where the administration kept us from getting into trouble."

Summer breaks were a time for outdoor adventures, jobs to earn spending money, and reconnecting with Valerie. At home in Montana, like his father and grandfather, Hood worked tough physical jobs in construction: digging ditches, laying pipe. One job for the Great Northern railroad involved elevating track embankments around curves to keep the trains on track. Under a blazing sun, men would lug two-hundred-pound railroad ties into place. "I learned more swear words than ever before in my life, and more words for intercourse," Hood said. "It was quite an education."

Hood also sought to engage his mind. Thanks to Caltech professor Ray Owen's suggestion, Hood visited Ernst Eichwald's lab in Great Falls. Eichwald was a pioneer in tissue transplant genetics and a former Harvard professor. Wearied of academic politics, he was drawn to Montana's outdoors. He hosted visiting scientists and was constantly on the lookout for bright apprentices. One was Irving Weissman, a high school student from Great Falls a year younger than Hood. Weissman didn't detect smugness from the Caltech student—just a hunger to learn. They hit it off, talking about their aspirations. Hood toyed with studying "psychobiology." It was a term then used to describe the biology of how the brain forms the human mind. Older scientists steered him away, as this was still science fiction. Psychobiology "was way too far ahead of its time," Weissman said. Many scientists would say the same thing sixty years later.

The summer before his senior year, before turning twenty-one, Hood did a grand tour of Europe with Adelberger and his brother. It was Hood's first time overseas. The students traveled by bicycle, getting deeply tanned. With barely enough money to get around, they stayed in hostels and, occasionally, tents. They wandered around Germany before exploring Switzerland, France, Denmark, the Netherlands, and England. Hood let his hair grow long. A couple weeks before returning to school, Hood wrote a searching letter to his sister, Doral:

> My family, Val, and school—each in turn—have
> come to mean far more to me than ever before.
> Yet, and this isn't inconsistent, you all seem so far
> removed. … let it suffice for me to say I seem very
> much to be living in a world far apart from the one I
> will come to know in ten days' time.

> Strange things have happened to this brother
> of yours. … you can't turn down an interesting
> conversation with someone from Great Britain or
> Canada or Australia, etc.—that looks at many things
> entirely different than you, and perhaps in a more
> mature way. Ah yes, I've learned a great deal this
> summer.

Life is a funny process, Doral. Just when you think
you've reached a new peak in "knowing" you get
a glimpse of what lies ahead and realize how little
of the road you've really traveled. … I'm afraid to
report, Doral, that in many respects, your brother is
still very selfish, but at least now he is working on
curing the disease much harder than ever before.

Hood closed by asking his sister for a Christmas present. He
didn't want a gift under the tree. He asked her to write a thousand-
word letter on what she wanted to do with her life. "Sis, this is no
joke—really that's what I would like more than anything else," he
wrote.

A year later, when he was starting medical school, he wrote
again on his sister's twenty-first birthday. He continued to wrestle
with questions of identity and his purpose in life:

It's strange, isn't it, how people always live for
the future—yet I'm certain some of our happiest
moments are in the past. Shelby, Missoula, and
beyond seem to lie so far in the past, but invariably,
my thoughts of that past are happy and carefree.
Do you remember moving to Shelby and the fear
and worry of meeting new friends and getting
adjusted—all of which turned out so well. With due
respect to the past, it's far more exciting to imagine
the future—and what it holds—love, marriage, a
family, and finally the wonderful feeling of knowing
you have given to this world in many ways. I would
suspect in the final analysis this is man's greatest
pleasure—knowing that he gave to that in which
he was interested, be it his or her kids, profession,

friends, or family. Sounds almost as if I'm leaning back to religion again—doesn't it—well, in some ways, I am.

He didn't lean very far, though. As a teenager developing his view of the world, Hood had shown little interest in religious traditions. Friends went to church, but religion was not a dominant influence in his community. Hood was baptized in a Christian ceremony as a toddler, but his parents didn't take him to church services. "My father was not at all religious; my mother was, faintly. She made a couple of attempts to get us to go to church. I just wasn't ever very interested," Hood said. "She wasn't insistent. We'd go at Christmas and special times."

In later years, Hood said organized religion is "good for many people," but like many biologists, he was agnostic. Even as a boy, Hood said, many parts of Scripture "didn't seem to make sense." He sought meaning through a scientific worldview—a belief in things that can be examined, measured, verified. The driving purpose in Hood's life, the thing he would pursue with missionary zeal in college and ever after, was science.

It wasn't that the giants of science on the Caltech faculty all exuded charisma and projected a contagious enthusiasm for their subjects. Despite the unusually tiny three-to-one ratio between students and faculty, undergraduates found it difficult to get close to professors who were focused on research. One of the giants Hood recalls giving lectures—Feynman—later acknowledged that his first try at teaching undergraduates was a flop. The material, as he presented it in the 1950s, was impenetrable for beginners. Feynman, one of the great characters in the history of science, came up with his popular "Feynman Lectures on Physics" later, after Hood graduated.

Key biology professors during Hood's undergraduate years included geneticists Ed Lewis and George Beadle, both future Nobel

Laureates. Lewis was a genius but "a pretty bad teacher," Hood recalls. His lectures focused on details of experiments. Similarly, Beadle tended to describe individual experiments, saying little about how they fit into big-picture concepts. It made genetics seem like a thicket of jargon. It was all trees, no forest.

Max Delbrück—a biologist whose quantitative rigor inspired James Watson—"was pretty terrifying to undergraduates," Hood said in his Caltech oral history. He could be "mercilessly caustic" when reviewing other people's work, wrote science journalist Victor McElheny. He often dismissed results with a trademark phrase: "I don't believe a word of it." Delbrück was a German-born scientist who originally trained in physics. "Here was this tiger of an intellect, who every year taught a different course. He taught the courses because he wanted to learn the material; I know of almost no one who's ever done that. He was probably a week or two ahead of the students in teaching these courses. But he had an enormous capacity for absorbing whatever it was," Hood said.

Years later, when Hood was a professor, his basement office in Caltech's Church Hall was next to Delbrück's. The two men had contrasting styles, with Hood pouring manic energy into broad visions that required big teams and big money. Delbrück kept his lab small, focused on problems that were narrow and deep. While Hood maneuvered himself toward the front edge of big waves in science, such as genomics, Delbrück was iconoclastic. Just as a crowd of scientists was moving into molecular biology in the 1950s, he switched his focus to an obscure type of fungus. He famously told young scientists: "Don't do fashionable science." That was not the kind of advice Hood would follow.

Although Hood was intimidated as an undergraduate by Delbrück, and the two never formally collaborated, Hood said he eventually grew fond of Delbrück. The molecular biology pioneer, in later years, became "much more of a human being. ... I enjoyed talking with him, and he always had wonderful insights, and he always pushed you right to the limits of your knowledge and understanding," Hood once said.

What drew Hood to biology? Two people, above all. His two favorite Caltech professors were father figures who communicated a sense of joy about science. They were James Bonner and Ray Owen. These teachers "helped me to appreciate the marvelous beauty and complexity of biology," Hood later wrote.

Bonner, a plant biochemist, took great delight in science. "His enthusiasm was wonderfully refreshing," Hood said. "In academic environments, we all have lots of work, and we tend to be overburdened, and we try to do too many things. There aren't too many people who have that enormous enthusiasm and passion and love for science. James really exhibited that in spades."

Owen, an immunologist, was an especially kind soul. Unlike other professors, Owen seemed to enjoy meeting freshmen and putting them at ease. He invited students to his house for dinner, and he encouraged them to tackle big problems. Immunology was in the dark ages then, and Owen made clear that the best was still to come. His signature contribution, in the 1940s, established that the immune system of cattle had "tolerance" or the ability to seek and destroy invading pathogens such as viruses and bacteria, while avoiding or "tolerating" healthy tissues. Hood was spellbound. Here was a vast frontier with significant consequences for human health. Further understanding promised to lead to effective organ transplantation and much more. Talks with Owen, over time, helped convince Hood to stick with biology. Most of all, Owen offered comfort. As Hood recalled:

> Ray was just one of these incredibly special people.
> It showed in every way. He cared about people. He
> had enormous faith in people. He had the ability to

draw out the very best in you. For many, many years, he taught this marvelous freshman course—topics in contemporary research, in which faculty members came to Ray's house. And Ray's personality presided over everything. They would give their talk about science and then they'd sit down and have cookies with the kids and talk about all sorts of things.

Hood was clearly seeking a father figure. In that sense, he was not alone. One psychological profiling study of forty elite scientists before World War II at Berkeley found that almost half were essentially fatherless as children, with "their fathers dying early, or working away from home, or remaining so aloof and nonsupportive that their sons scarcely knew them." Many of the Berkeley scientists weren't necessarily drawn to science by the pleasure of learning things so much as they were on a journey of personal discovery. "The fatherless young man finds a masterful surrogate father of warmth and dignity, identifies with him and proceeds to emulate him," wrote the author Richard Rhodes. By his junior year of college, Hood settled on a career in biology. Hood's connection with Owen was so strong that the two men maintained a scientific correspondence into Owen's nineties.

The late 1950s were heady days in biology. The discipline of molecular biology was just emerging. For the first time, scientists were able to answer questions by diving deep into the underlying code of DNA (deoxyribonucleic acid), RNA (ribonucleic acid) transcripts, and the large protein molecules that do most of the fundamental work in cells. Before James Watson and Francis Crick's discovery of the DNA double helix in 1953, biology was ruled primarily by biochemists, who studied amino acids and larger structures within a cell. Even after the discovery, many traditional biochemists held firm that their approach reigned supreme.

But a revolution was brewing, as scientists realized that there was a unifying code of DNA present in all forms of life on earth—plants, animals, bacteria, humans. That code carried the information, a set of instructions, for making the heart cells that form heart tissue, the kidney cells that form kidneys, and so on. However, biologists didn't yet have a way to precisely and reproducibly measure basic facts of nature, the way physicists could with, say, the speed of light. Natural organisms seemed to adapt and mutate in strange and unpredictable ways. Especially in areas such as genetics and immunology, it was difficult to even collect consistent quantifiable measurements or settle on consistent definitions for a "gene." The field was qualitative and unfocused. Delbrück, highly influential, argued forcefully for quantitative measurement. In the late 1950s, many smart people believed biology was on the brink of big things. Young people followed.

This was the environment in which Hood embarked on his path in biology. Scientists from other fields came, too. Haunted by the atomic bomb after World War II, many physicists sought new challenges in biology. Delbrück, Leo Szilard, Salvador Luria, and Francis Crick were just a few physicists who made the switch.

Caltech's biologists focused mostly on plants, insects, and viruses. These were logical places to start. Those organisms tended to be easier to investigate than monkeys or humans, and they came with fewer ethical restrictions. But as it became clear that DNA code ran through all forms of life, many young scientists were intrigued to expand the field of study. Here, in this nascent discipline, lay the recipes for making proteins, organs, and, ultimately, complex higher organisms.

Hood was hooked.

He graduated with a bachelor's of science from Caltech, with honors, on June 10, 1960. He got a 3.6 grade point average on a 4.0 scale. He ranked No. 7 in a class of 147. As a senior, he received the Hinrichs Memorial Award, given every year to a Caltech undergraduate who made the greatest contribution to the student body and "whose qualities of character, leadership and responsibility have been outstanding." It was a great achievement for the small-town kid. Still, he had more to learn.

Hood had little interest in plants, viruses, bacteria, or model organisms like *C. elegans*—worms. He wanted to see what made humans tick. Given Caltech's slim offerings of human biology in the early 1960s, Hood looked elsewhere for graduate training. Medical school was one place to learn the basics, and Caltech didn't have one.

The young man saw a new frontier. It was time to venture east.

Chapter 5

The Making
of a Young Biologist

One of the big, elusive questions in immunology was: How does the body produce so many billions of antibodies, each one structurally distinct, to guard against the infinite number of viruses, bacteria, pollens, and other substances that people encounter every day? During a microbiology class in his first year of medical school, Hood became fascinated by the question.

Hood had applied to two of the best medical schools in the world—Harvard Medical School and Johns Hopkins School of Medicine. Both accepted him. At first, Hood planned to go to medical school for two years, then transfer to a PhD program in biology. Caltech's James Bonner encouraged him to pick Johns Hopkins in Baltimore because it offered an accelerated, three-year program. That was all a young man in a hurry needed to hear. Hood resolved to use his MD as a stepping-stone to the PhD he needed to become a researcher. This was before medical schools began offering concurrent MD-PhD programs for people who wanted training in research and patient care.

Hood's microbiology class was taught by a famous physician-scientist—and a former Harvard quarterback—William "Barry" Wood Jr. Wood discussed the antibody mystery.

Scientists knew that the body didn't have billions of unique genes for making a specific antibody to neutralize each invader (known in immunology as an antigen). Humans had to make billions of antibodies to fight off billions of antigens in some other way.

How?

There were a couple of competing hypotheses in the early 1960s. Some scientists were betting that the immune system used invading antigens as templates—that antibodies adjusted to their shape and folded around them. Another school of thought held that the body had several different sets of genes, which could be summoned to make unique antibodies to ward off unique invaders. Given the state of technology, the answer was unknowable in the 1960s. But the seed of a powerful idea had been planted.

"It was an intellectually exciting question," Hood said of the underlying basis of antibody diversity. The question would drive Hood's career, off and on, for more than twenty years.

Caltech had prepared Hood well for medical school—and the rigors of molecular biology. Sloppy thinking wasn't tolerated. Sweeping conclusions had better be grounded in hard data. Experiments had to be well designed to limit variables and reduce potentially confounding factors. Woe be it to those who leaned on artifacts—accidental effects that prompt faulty conclusions. "One of the great things about Caltech was having peers who set your standards and expectations," Hood said. Once, when debating classmates about the cause of diarrhea, the other students pointed to microorganisms. To Hood, coming from a rigorous and quantitative educational background at Caltech, that was a superficial answer he refused to accept. He wanted to know the underlying patho-physiologic mechanisms causing diarrhea once the microorganism invaded. Hood was already thinking more like a scientist, rather than a physician. Only one of the most senior clinicians they asked knew the answer at that deeper level.

"That's what's wrong with medical school education," Hood said in his Caltech oral history. "You had so much to learn that you were exhausted to the point of not being able to even try to think about things and put them in context."

The workload was grueling. Hood was used to that, too. Hood went through the usual clinical rotations, hanging around surgeries, emergency room duty, and women giving birth. He performed well at Johns Hopkins. Despite a few Cs (in genetics, physiology, surgery, and public health), he was elected to the prestigious Alpha Omega Alpha honor society. By the end of medical school, the department chairman urged him to stick with medicine and not waste his training.

Hood wasn't persuaded. Treating patients every day struck him as repetitive and dull. "I really liked interactions with patients. I liked problem solving. But it was clear to me that once you went through and learned the high-level stuff, then 95 percent of it would be the same stuff over and over again. The 5 percent was what the good doctors really love," Hood said. The best physicians also have great bedside manner, a deep sense of empathy. That wasn't his strong suit. "The first time I ever saw a patient, it was this eighty-two-year-old Jewish woman who was tough as nails," Hood said. "I walked in, and she said, 'You're no doctor. You're a medical student!'" Instead of acknowledging her feelings, he thought she had some nerve. He stood his ground, saying he was a *young* doctor perfectly capable of handling her case. The patient was unmoved. "I want a *real* doctor!" the woman said.

Hood talked science late at night with a friend, Steve Rosenberg, often at the cheapest Chinese restaurant they could find. (Baltimore didn't have any mountains.) The action, they agreed, was not in the clinic, but in the lab. "We very quickly became close friends," recalled Rosenberg, who later became a distinguished scientist at the National Cancer Institute. "We were in medical school because we wanted to develop the medicine of tomorrow."

Feeling the itch for research, Hood worked briefly in the lab of neurophysiologist Vernon Mountcastle. Hood's passion was crafting theories and testing them in experiments. But he was too low on the totem pole and didn't know enough to be crafting theory. Instead, his job was to record nerve impulses. The research seemed to have more to do with electronics than biology, and his role wasn't far above cleaning rat cages. "It made me wonder whether basic research really was where I should be," Hood later told a magazine writer. "I viewed myself as a failure."

But he did find the technology interesting. Intrigued by the mystery of the immune system, Hood believed new technology would someday enable him to attack such a problem—the sort of thing that won people a Nobel Prize. But the glory in science went to those who made discoveries. Machines were best left to engineers. Most scientists would just try to develop the best theories, and the best experiments, with the tools they had.

Medical school gave Hood more than the grounding in human biology that he sought. He gained some sense for what a physician encounters when treating patients. He could speak their language. These things helped him create a rapport with physicians—critical gatekeepers to patients and the samples they provide that are the lifeblood of research. Plus, he learned a musical instrument. Hood's girlfriend, Valerie, had graduated from Whitman College and moved to Washington, DC, to pursue a master's degree at George Washington University—just thirty-three miles away. Hood drove from Baltimore many weekends to visit her. Sitting in traffic was deadly dull. To pass time, he played the harmonica.

Hood was twenty-five and in his final year of medical school. He needed to decide where to pursue his PhD, and whom to hunt down as his thesis adviser. As an undergraduate and medical student—seven years of prerequisite book learning—Hood had had little time to do what he really wanted to do—scientific experiments. He kept his finger on the pulse by reading the scientific literature. Now it was time to become a real scientist. He wanted to be someone who advanced understanding in human biology, who broke new ground.

He applied to at least six graduate schools, including Harvard, Caltech, and the Massachusetts Institute of Technology. MIT wooed him, but the school didn't have anybody working in his specialized area. Caltech was familiar, but should he go somewhere new for a different perspective on how science was done? Hood had been reading about a thirty-five-year-old scientist named William J. Dreyer. Working at the National Institutes of Health along with Mike Potter, Dreyer had developed a system to characterize and evaluate differences among antibodies in mice with myeloma, a form of cancer. This was one potentially useful piece for figuring out how the immune system was able to create such a diverse range of antibodies—"antibody diversity." The year Hood graduated from medical school, Dreyer left the NIH for a professorship at Caltech.

Hood contacted Dreyer and said he wanted to work with him. Dreyer believed that with superior technology, gathering different kinds of data, he could beat other scientists to the punch. Hood was intrigued. Dreyer invited Hood to be his first graduate student at Caltech in the fall of 1963. That settled things. Graduate school would be back in his old stomping grounds.

Valerie had made a decision as well, after dating Hood off and on for almost eight years. "I have something important to tell you,"

Valerie said. "I have a really nice job offer in Lebanon, and I'm going to take it unless you marry me." Hood proposed.

Leroy Hood and Valerie Logan were married in a small ceremony on a clear Saturday afternoon, December 14, 1963, in Pasadena, California. The guests were mostly graduate school friends. The decision to get married far away from Montana didn't sit well with their families. The couple agreed to hold a second ceremony for extended family members during Christmas break. On December 29, at the First Presbyterian Church in Logan's hometown of Cut Bank, Montana, the two repeated their vows. Hood's younger brother Myron was the best man. The minister was "really unhappy" about not performing the original ceremony, Hood said. "I tried to mitigate his annoyance by talking football." After the reception, the newlyweds visited Red Lodge—hometown of Roy Wadsworth, the grandfather Hood considered a true father figure.

The couple's wedding photo shows a smiling Hood with a military-style buzz haircut. His chin is down, forehead tilted forward, eyes focusing ahead as if to burn a hole in the camera. Even on a happy occasion, his posture was tightly wound, cobra-like. The newspaper caption below the wedding photo announced "Mr. and Mrs. Leroy Hood."

Hood and Valerie settled into a small apartment on North Holliston Avenue, about four blocks from Caltech. Valerie took a job as a high school teacher. Graduate school at Caltech was not as intense as undergraduate school. Competition was less fierce, as most courses were graded on a pass/fail basis. But, as always, Hood set his own level of intensity. He kept eighteen-hour workdays. Viewing physical endurance as one more way to gain a competitive edge over other scientists, he ran a few miles and did one hundred push-ups every morning before class and lab work. Classmate Regis Kelly, now director of the California Institute for Quantitative Biosciences, remembers asking Hood to loosen up once by playing rugby. No dice. "He was a bit of a straight arrow," Kelly said.

Hood was usually still in the lab late into the night. If his bride wanted to spend time with him, she tracked him down there. Valerie would often join her husband during nighttime coffee breaks with colleagues. They would talk about current affairs in politics, culture, and the grind of graduate student life. The group held weekly seminars to debate the importance of new scientific findings. "There was a whole night-lab culture," Valerie said years later. A few old friends were still around, including his former roommate, Eric Adelberger, who was then a Caltech graduate student in physics. They resumed their weekend camping and hiking. It was Hood's favorite way to relieve stress and recharge his batteries. Valerie, a petite woman of little more than a hundred pounds, would often shoulder a heavy backpack and join Hood on trips to the Sierras and the Grand Canyon.

Hood may have done a little soul searching in the letter to his sister about being self-centered, but he hadn't changed much. One time, he planned a week-long backcountry camping trip for himself and Valerie in Sequoia National Park. His plan was to hike twenty miles a day in high-altitude terrain—more than twice the distance of the average backpacker. The trail was wickedly steep, climbing about four thousand feet in just the first two miles. He and Valerie were carrying a hundred pounds of gear, food, and water between them. Hood, still in football shape, galloped along, enjoying the scenery. He soon realized his mistake. About two-thirds of the way, Valerie yelled out, struggling to keep up. The trip was cut short. Hood sometimes told the story years later to poke fun at his own zealous approach to the outdoors, but Valerie wasn't amused. "I didn't really think of her," Hood said.

One of Hood's main tasks in graduate school was to learn how to think like a great scientist. He wanted to think about attacking big problems, not small stuff. Graduate school, Hood realized, wasn't just about doing an obscure research project to earn a PhD. It was about figuring out the habits required for a successful career in science. "My view of graduate school is there are two things you really need to learn," Hood said. The first is "to learn how to read the literature and realize that not everything you read is true. The second is to learn how to think about experiments and designing experiments—thinking in a bigger context about what the major problems are." With his independent spirit, Hood thrived. "I did both of those things at Caltech very well," he said.

As Hood's thesis adviser, Dreyer helped him think big. Dreyer had a special ability to think about molecular interactions in vivid mental images—and in broad conceptual strokes, way beyond the body of scientific evidence at the time. Even when going out on a conceptual limb, he didn't suffer from self-doubt. He was unafraid to play the role of the maverick, or even, in his own words, the "kook."

Throughout his career, in speeches and interviews, Hood consistently credited Dreyer as his most important influence in science. In an interview in 2012, eight years after Dreyer died, Hood recalled:

> Bill was a very complicated guy. I was never quite
> sure of all of his motivations. He was inspirational.
> He gave me two things that I will always value.
> One of the things he said was that if you're going
> to do something in biology, be at the leading edge.
> It's much more fun there. The second was that if
> you really want to change a field, invent a new
> technology. Those are things I took quite seriously
> for my entire career.

Dreyer saw a bright and driven young man with potential, and one who appreciated his conceptual, visionary, "don't worry about the details" style. Dreyer's small lab when Hood arrived was primarily devoted to determining the structure and function of lysozymes—enzymes that play a role in fighting infection by attacking bacterial cell walls. A couple of postdocs and technicians plugged away at identifying the amino acid sequences that could help them deduce the answer.

Hood was eager for a piece of what he saw as bigger action. That was the work Dreyer had begun at the NIH, and continued with collaborators there, on studying antibody diversity in mice. Dreyer said Hood could pursue the work as a low-priority "Saturday afternoon" project when all other regular duties were completed. "It was exactly what I wanted," Hood said.

The mentor, however, didn't provide much mentoring. Dreyer's management style could best be described as "hire good people, inspire them, and get out of the way." It was a successful model for self-motivated students like Hood, but it could be taken to extremes. Dreyer could get so far out of the way that people might be left wondering where he'd gone. (Years later, Hood adopted that style himself.) "Bill was never interested in the details of anything," Hood said. "He liked to think about higher-level things. He was fun to talk to about big ideas." (Hood's graduate students and postdocs said the same thing, over and over, about him.) So the nuts and bolts of running successful experiments were left to students. Hood would start by reading the literature on antibody diversity.

Antibodies, as any beginning graduate student knew, are the large, Y-shaped proteins that can bind with specific structures on invading pathogens, or on specific structures that appear on the surface of cells. Nature could make such an amazing array of antibodies that immunologists would often say it was like making specific keys for specific locks. Not only could the immune system make keys for locks, but also it was capable of adapting to make keys for almost any kind of lock. The breadth of the immune system's repertoire was astonishing. Many different types of cells act in concert during a fully functioning immune response. When a foreign virus or bacteria invades the body, B cells (so-called because they're produced in the bone marrow) are alerted to secrete antibodies. But that's just one small part of a bewilderingly complex response. Signals are sent to many different types of white blood cells, which then go through an intricate dance of alerting and recruiting other cells, which then swarm the invader and kill it. A big part of scientists' challenge was to break down this symphony—with billions of cells and proteins doing different things—and isolate the specific role of each musician in this score.

Where to start? Mice. Hood settled on a specific, narrow problem. One type of cancer of the bone marrow—multiple myeloma—consistently threw off telltale cancer proteins in the urine of mice. Those proteins, which can be extracted and purified, are called Bence-Jones proteins. Back then, scientists couldn't run those samples through a machine to determine the underlying DNA sequence. The technology didn't exist.

The work of getting the proteins and determining their sequence was painstaking. For starters, the samples were tiny. Mice are small; they urinate in small volumes. Each small sample held only trace amounts of the telltale protein Hood was looking for. Once Hood had gotten a few precious milligrams of the protein, disaster could easily strike. If even the tiniest amount of oxygen got into the experimental chemicals, it could wreck weeks' worth of research. During graduate school, he often stayed deep into the night doing this painstaking manual work. Great care had to be taken to run these experiments right.

Once he had isolated the precious Bence-Jones proteins, Hood would be able to determine the order of the amino acid building blocks that made up the proteins. Protein sequencing was primitive then, but possible. With that information, he could work backward and deduce the underlying genetic code. If he learned the order of amino acids in about a dozen Bence-Jones proteins, he could compare the amino acid sequences to see what parts were the same and what parts were uniquely arranged to encounter an invader.

Self-doubt, that constant companion of graduate students toiling in the night, haunted Hood. "He was, like all first-year grad students, kind of a klutz, working very hard," Dreyer recalled. "He was very motivated, and a very competitive guy, but he didn't get an awful lot done that first year."

"Hampered in his grand plans by somewhat cumbersome techniques," one magazine later wrote, "he asked himself whether he would ever make a good experimentalist." As a biologist trying to understand theory, Hood didn't know enough about the cumbersome instruments to reengineer them himself. Unlike his father, Hood was not a tinkerer. But he was more certain than ever that biology needed better technology to automate the most tedious experimental processes.

Two years into Hood's "Saturday afternoon" project, Dreyer dropped a bombshell. His lysozyme project wasn't working out. He was taking over Hood's project and having a senior postdoc sequence the proteins Hood isolated. "Bill and I had an enormous battle," Hood said. "I thought it was entirely unfair. It was my project. I had done all the work. How could he do this to me?"

Dreyer was the boss. In the feudalistic world of academic science, graduate students—the peasants of the enterprise—regularly get trampled. Professors take credit for work that occurs in their labs. Dreyer put senior postdoc J. Claude Bennett on the project. Hood joined them for many conversations about the hypothesis and how the experiments being done might advance the idea or debunk it.

But he didn't get to finish the work that would put his name at the top of the published paper. "Dreyer said, 'Look, Lee, you're very good, you'll figure out other interesting things to do,' Hood recalled. "Part of me never really forgave him for it."

Hood's project was related to Dreyer's view—heretical at the time—that antibody diversity is possible because parts of genes reshuffle themselves when B-cells form antibodies. Dreyer's notion contradicted a theory advanced by Nobel Laureate George Beadle. Affectionately known to colleagues across biology as "Beets," he was known for his intelligence, his sense of fairness, and a lack of arrogance. The former Caltech biologist was serving as the president of the University of Chicago in the 1960s. Scientifically, he was most famous for a seminal discovery showing that distinct genes provide the instructions for making distinct enzymes, which, in turn do much of the work in cells. It was a big deal: he and collaborator Ed Tatum were the first to show what genes do. Their work was later summed up in a catchy slogan: "one gene, one enzyme." People just didn't go around casually challenging the work of the great George Beadle. End of story.

Beadle's insight was correct in some respects, but by the 1960s, Dreyer thought there was more to the story. At least in the case of the immune system. Dreyer maintained a collaboration with the NIH on studying myeloma tumors from mice. The mice could provide crucial samples and data to test the hypothesis. Together, Dreyer and Bennett published a provocative paper titled "The Molecular Basis of Antibody Formation: A Paradox," which appeared in the *Proceedings of the National Academy of Science* in 1965.

In that paper, Dreyer and Bennett suggested—way ahead of their time—that antibodies achieved such diverse adaptive ability because they could be formed with DNA from two genes, not just one, as most scientists believed. Like Beadle's "one gene, one enzyme" insight years earlier, Dreyer's could be reduced to a slogan: "Two genes, one protein." Turning a complex idea into a catchy one-liner was a clever technique some scientists used to spread the influence of their ideas. It became one of the essential skills Hood used throughout his career when communicating with insiders and lay audiences.

In the case of "two genes, one protein"—or "two genes, one polypeptide chain" as it was sometimes called in scientific circles—the idea was that one gene was responsible for recognizing the foreign substance to attack and another was responsible for the killing function of the antibody produced by the B cells. The two genes would be physically rearranged and combined as the B cells matured. Instead of a finite number of genes laying out instructions for making a finite number of antibodies, the possibilities for making antibodies through this reshuffling of genetic information could be infinite.

There was only one problem: Dreyer and Bennett didn't have enough data to provide convincing evidence of such an amazing biological phenomenon.

Reaction was swift and fierce. As Dreyer recalled of his immunology colleagues:

> They just didn't see how this could possibly be true. I had to ask myself why that was so, and part of it is the old [story about what happens] when someone from the outside comes in to a field where everybody believes a different set of beliefs, like a religion.
>
> I was the guy from the outside who said, "This is crazy. This is the way it is, really." So why was it so hard for Ray Owen and others to understand this? It violated the paradigm.

Dreyer wouldn't back down. Hood, as a student, didn't see a rearrangement pattern at first. "Lee didn't get it quite, just because he didn't have the experience," Dreyer recalled. "Plus, he's never been—he's a different kind of person. He's superb as a scientist, and he's superb as an administrator and fund-raiser, but he wasn't a risk taker, or an imaginative innovator, or whatever. And I am to a fault, so we make a good pair. We still do, to this day."

But as the experimental data mounted over a year, Hood saw that his adviser was right. The immunology community was clinging to outdated dogma. The intellectual debate created some serious fault lines in Caltech biology. Owen, the biology chairman, received considerable grief from colleagues around the country who asked how he could allow Dreyer to publish such nonsense challenging Beadle's work. Owen himself thought the antibody diversity concept was "bullshit," in Dreyer's words. The chairman went so far as to block Dreyer from getting any funds to develop instrumentation that might prove the theory. Owen backed down—allowing Dreyer some modest engineering funds—only when Dreyer threatened to leave Caltech.

The experience taught Hood a lesson that stayed with him his entire career. Scientists, those men and women who are supposed to live in a world of possibilities and utter data-driven rationality, can dig in their heels and be as dogmatic as religious zealots. "I realized for the first time how threatening new ideas are to many scientists," Hood said.

Controversy didn't scare off Hood. He still stewed with outrage over losing the project that had been his for two years. He had done a lot of hands-on work and had been part of lively conversations with Dreyer and Bennett on how to interpret their data. Still just a lowly graduate student, he felt cheated out of his fair share of the credit.

> Bill [Dreyer] came to me and said, "I don't think you
> should be a named author on this paper, because
> it's really going to be controversial, and it might
> not be good for your career." I never believed that

for a second. But that was it. I wasn't on the paper.
One other postdoc and he were on the paper. I have
to tell you, it created a firestorm. ... I had been
involved in a lot of the discussions, as much as the
other postdoc had been. So yeah, I felt I deserved to
be [an author on the paper].

Even though Hood didn't have the first thunderbolt of insight
about the two genes rearranging, he had an uncanny ability to recognize a good, but raw, idea when he saw it. As he did many times
throughout his career, Hood got on board with a promising original
idea, and became an early and forceful advocate for it. Hood fully
embraced Dreyer's provocative new theory of antibody diversity.

Dreyer recognized Hood's driving ambition and sought to stoke
the fire. As a relatively green second-year graduate student, Hood
was given the rare privilege of presenting results to peers at other
top institutions such as UCLA and the Salk Institute in San Diego.
The young scientist, armed with intriguing data and charismatic
public speaking skills honed in high school debate, relished the
intellectual joust with the skeptical older men (they were mostly
men) of biology.

Working on such a high-profile problem and generating provocative results put Hood on the fast track in this exciting application of
protein chemistry with far-reaching implications for immunology.
Debate raged about the Dreyer lab's findings. Dreyer was invited
to discuss the results at a June 1967 symposium organized by Cold
Spring Harbor Laboratory, an elite research haven on New York's
Long Island. The biggest guns in biology gathered, including Nobel
Laureate Francis Crick and two future Nobel winners, César Milstein
and Gerald Edelman. Hood, a few months shy of his twenty-ninth
birthday and still a year away from his doctorate, was thrilled to rub
elbows with the scientific elite.

Dreyer and his student presented more data to support their theory, but they still hadn't completely nailed it on the big stage of Cold Spring Harbor Laboratory. Fred Sanger's famous invention of a gene sequencing method—a way to read the informational code carried by the double helix—was still a decade away from widespread use. There was no practical way to show definitive proof that DNA sequences were being reshuffled from two genes to form one antibody.

Hood and his mentor thrived on the uncertainty. They acknowledged the limits of their data, without backing away one inch from the big idea. In the opening of their paper presented at Cold Spring Harbor, titled "The Genetic, Molecular, and Cellular Basis of Antibody Formation: Some Facts and a Unifying Hypothesis," the scientists went so far as to borrow a quote from one of Abraham Lincoln's addresses to Congress. "The dogmas of the past are inadequate to the stormy present ... as our case is new, so must we think anew and act anew. We must disenthrall ourselves," they wrote. Even though the hypothesis was based on a thin set of antibody data, they defended it with gusto. The problem, as they saw it, was that protein sequencing was so primitive, there was no way to get all the data needed to prove them right.

The clash of egos and the competition was thrilling to Hood. It reminded him of big-game football intensity. As he recalled in his Caltech oral history:

> It was very heady stuff for a second-year graduate
> student to be going to these places and talking
> about these kinds of things and being asked to
> major meetings on antibody diversity. ... Within
> the year, this whole field exploded. There was a very
> famous workshop at Warner Springs [California],
> where people from Rockefeller University and from

Indiana University, working in similar areas, got in an enormous fight. Each group had enough information, which, if you put it all together, would come out with a really interesting picture. But neither wanted to give up the information so that the other could formally prove this two-gene, one-polypeptide chain hypothesis. And it was then that I realized ... this was no Saturday afternoon project. These were the big boys.

Although Hood felt that he had been elbowed out of an important paper a couple years earlier and that his work was underappreciated, Dreyer disagreed. He said he was being charitable in giving credit to a graduate student who didn't see the whole picture and didn't have the key theoretical insight in the first place. Dreyer, in his oral history, said he purposely allowed Hood to present the findings on the main stage at Cold Spring Harbor, sharing the big stage, and letting the young man get a piece of the credit. Dreyer knew that would motivate a student who craved the spotlight. Thirty years later, Dreyer said:

> If you want a protégé to take off and get fired up and work hard, you let the protégé take credit. And that's Lee Hood. He was a very gifted student, very competitive, and very hardworking. He only sleeps three to four hours a night ... these guys—even Thomas Edison—had catnaps. I do that. It may have to do with the energy required for this enormous computing power of a Silicon Graphics–like brain.

People who worked with Hood throughout his career would often chuckle about those "catnaps." Often, Hood would nod off while listening to scientific presentations. Then, suddenly, he'd spring awake and ask the speaker a question that cut to the heart of the matter. It was as if he heard every word.

By the summer of 1967, Hood had done the work he needed to get his PhD. But writing his thesis and defending it went on the back burner. The Vietnam War was heating up. Hood, like many educated men who were drafted, sought to fulfill his service duty without getting shot.

Hood flirted with taking a research job at the University of Oxford in Great Britain, but he wasn't prepared to face the consequences of draft dodging. Hood faced his draft board. By 1967, the board had grown quite accustomed to hearing excuses. Hood tried to argue that although he had a medical degree, he wasn't a board-certified physician and therefore wasn't qualified to treat wounded soldiers. No problem, he was told. He could still take medical histories and do basic physical exams on the battlefield. However, with medical training, Hood did have three attractive alternatives for serving their country in wartime. He could go to an Indian reservation, to the Centers for Disease Control and Prevention, or to the National Institutes of Health's Public Health Service. The NIH campus in Bethesda was a research mecca. "Going to NIH was one of the simplest decisions I ever made," Hood said.

Usually, a graduate student writes and defends a doctoral thesis, gets a PhD, and takes a postdoctoral fellowship. The fellowship is a time-honored rite of passage, long on hours and short on glory, with starvation wages. Standout postdocs, with sacrifice and a little luck, use the fellowship as the springboard to a coveted faculty gig.

Hood figured he'd skip that part. With more than a little chutzpah, Hood told an institute within the NIH that he wanted a senior investigator position. That meant he'd get a higher salary than a postdoc—plus his own budget, and his first staff of postdocs and technicians to supervise. John Fahey, head of immunology at the National Cancer Institute, knew Hood was an up-and-comer in the hot field of protein chemistry. "[Fahey] said 'If you come to my branch, I'll give you all the money you need to set up protein chemistry here.' It was a permanent position. So I said, 'OK, sounds great,'" Hood recalled.

He spent his first few months at the NIH finishing up the "dreary" writing of his PhD thesis. It carried the rather dull title of "Immunoglobulins: Structure, Genetics, and Evolution." He got the all-important degree in June 1968. It was a relief. Hood dedicated his doctoral thesis to "the Grand Canyon country and the Sierra Nevada, for the constant renewal and release they have provided these past four years—to my wife, Valerie, who shared my wilderness pleasures—and to all who fight to preserve the mountains and the canyons for future generations." Although Hood thanked Dreyer for the "broad horizons which he provided," he was still irked at the boss. Dreyer didn't bother to provide feedback on the dissertation. Hood wasn't sure he even read it. Instead, an old friend on the biology faculty at Caltech, Ray Owen, helped Hood finish.

Owen, Hood's favorite old professor and now the chairman of biology at Caltech, encouraged Hood to keep pushing himself at the NIH. In keeping with academic tradition, it would be best for Hood to escape his adviser's shadow and prove he could do good work on his own. Do that, and doors might open again someday on the Caltech faculty, Owen said.

The NIH experience was better than either man imagined. Up until this point, Hood had mainly toiled in obscure labs, late at night, with limited opportunities to build relationships with scientific leaders. Bethesda, by contrast, was loaded with big scientific names. Ordinarily, bright young scientists would be scattered among university campuses, but because of the war, Bethesda had become a magnet for them. Christian Anfinsen, a legendary biochemist who won a Nobel and inspired a generation of scientists, was there. So were Julius Axelrod, Marshall Nirenberg, and Gordon Tomkins. A younger generation that included Bill Rutter, Joseph Goldstein, Michael Brown, Anthony Fauci, and Gilbert Omenn joined Hood among the up-and-comers. Several of the younger scientists jokingly called themselves the "Yellow Berets," in contrast to the battle-tested Green Berets of military legend. They didn't try to deny they were there to avoid the rice paddies of Vietnam. Hood's contemporaries there went on to run medical centers at Stanford University, Massachusetts General Hospital, and other top institutions. "In that period of time, NIH was better than any place," said Omenn, a University of Michigan biologist who became a lifelong friend of Hood's.

By talking to leaders across multiple disciplines and absorbing many big ideas in science, Hood honed his intuitive sense for where biology was heading. He decided not to let himself get pigeonholed in a narrow discipline. Years later, colleagues spoke of Hood's uncanny knack for identifying the best ideas of others, and his excellent taste in selecting high-impact projects for himself. "What makes a really good scientist great is the intuition to choose your path to optimize the yield of what you can learn, and move in directions that are most fruitful. It's not something easily taught," Hood said.

Hood also got experience in management and motivating others. The government had ample resources then. He had as many as five staffers working for him. This meant Hood, not yet thirty years old, would no longer toil at the lab bench. Others could squeeze pipettes into petri dishes. From now on, labor could be delegated. Besides, he had so many ideas to pursue, Hood would never be able to do everything with his own two hands. His new role enabled him to hedge his bets on many projects, as many experiments naturally failed. "I think I was very good at the bench. I enjoyed doing lab work, but I also enjoyed a broader scope and being able to attack problems in a broader way," Hood said.

He also was home in time for dinner. That was new. (It didn't last.) He and Valerie started a family. A son, Eran, was born in 1968. A daughter, Marqui, followed in 1970. Like many women of her generation, Valerie stayed home for the next five to six years to raise the children. Like many men of his generation, Hood was free to pursue his career full-throttle because of his supportive, stay-at-home wife. He didn't change his share of the diapers or do his half of the household chores. Hood vowed to be a more engaged and loving father than his own dad, but "Valerie was the primary caregiver, no question," he said.

Marqui was named after one of Valerie's college friends. Eran was named to honor a Jewish family friend, Henry Metzger, who had a son by that name. While most people in the United States spell the name *Aaron*, Hood went with the Hebrew spelling. "I really like the Jewish people. There's a part of me that really resonates with them," Hood said. It was an unusual comment for a man who never met a Jew until college. But he could relate to feeling like an outsider. Later in life, he traveled frequently to Israel to give scientific talks and negotiate partnerships.

For now, Hood was growing restless. By 1970, his service obligation was fulfilled. Hood craved action. He wanted to leave a mark on history, much like those famous faculty members he worked under at Caltech, Johns Hopkins, and the NIH. He was beginning to generate a lot of ideas, which were beginning to coalesce into a worldview. A large bureaucracy like the NIH was no place for an entrepreneurial spirit determined to challenge the status quo. The NIH had abundant resources and outstanding people, which Hood appreciated, but it was also a government agency. That meant slow, consensus-driven committees had considerable power. Hood didn't want anything that could slow him down. He wanted to ask and answer his own research questions. He needed a stage on which to perform.

At thirty-two, Hood was sure good things would come his direction. His bread-and-butter discipline, protein chemistry, was hot and getting hotter. He had experimental lab skills. He had a solid theoretical grounding in immunology. He had the drive to excel. He had an unusually high-powered network. He had learned about the culture of science: how harsh it could be for the little guys, and how much better it was to be a professor—a member of the nobility. And he had learned that imagination, or pure creative genius of the "Eureka" variety, wasn't the only path to a breakthrough idea. Hood knew there were other ways, more methodical ways.

As Nobel Laureate Albert Szent-Györgi once said, "Discovery consists of seeing what everybody has seen and thinking what nobody else has thought." Sure, Hood could spend his career trying to come up with new experiments by using the same tools everybody else had. Or he could follow Dreyer's advice: shake up the field by inventing a technology that throws off all kinds of new ideas and tears down old ones.

A plan was beginning to come into focus. Hood would gun for a faculty position. He would chart his own course. He would do it at the intersection of two worlds—biology and technology.

Chapter 6

A Rising Star for the Biotech Era

"Let me tell you what this place is really like," a friend at Harvard Medical School told Hood. "My lab is smaller than the bathroom in the chairman's office."

Hood was enjoying life as a free agent in the spring of 1970. He flirted with joining the faculty at several leading universities, including MIT, Stanford, Caltech, and—until those discouraging words—Harvard. "It was a bit of an exaggeration, but he did make the point about the classes of citizens at Harvard," Hood said. "He said he wanted out. It was a very hierarchical place." Three years later, the friend, Mario Capecchi, took a faculty job at the University of Utah. Capecchi spread his wings there and went on to win the Nobel Prize for devising ways to turn off certain genes in mice. The "knockout mouse" technique became a widely used enabling technology, helping biologists understand the function of individual genes.

Brimming with ambition in his early thirties, Hood wanted to be someplace where he could set his own research agenda and advance on the merits. He knew he wanted to work at the nexus of biology and technology, but he didn't yet have a fully formed idea of what that meant. At a minimum, he wanted to be somewhere he could figure it out without waiting his turn or being anybody's deputy.

He looked carefully at three places in the West: Stanford University, the University of Denver, and Caltech. Stanford was an intriguing possibility. Not only did it have an excellent reputation, but old pal Irv Weissman had recently joined the faculty.

Stanford offered Hood a faculty job at a starting salary of about $20,000. It was generous for 1970—the equivalent of $122,000 in 2014 dollars. Still, Hood felt the pull of Caltech. It was familiar: Hood had a sense for who had power, who didn't, and who was allied with whom. The place wasn't excessively hierarchical. Its excellence was undeniable. Friends were close. So were his beloved mountains. On the down side, Caltech biology was dominated by an older generation of scientists, few of whom he had bonded with. He also had a complicated relationship with Bill Dreyer, his graduate adviser.

Caltech, for its part, saw the upside. Many people there saw Hood's potential. Ray Owen, the Caltech immunologist, had told Hood that if he did good work elsewhere, he might be welcomed back to the faculty. "Caltech biology was a bit of a sleepy place. ... It was a department that was just beginning to think about revitalization—bringing in a younger set of people who would be dominant figures," Hood said. When word got back to Pasadena that Hood was interviewing for faculty positions at rival institutions, the chairman of biology, Robert Sinsheimer, pounced. He offered $16,000, the equivalent of $98,000 in 2014. When Hood asked whether Caltech would match Stanford's offer, Sinsheimer said no. "I guess you'll have to decide what's most important to you," the chairman said. Hood accepted Caltech's offer.

Here was a place where Hood believed he could fulfill his dreams. Caltech was a magnet not only for gifted science students but also for engineers who could apply scientific knowledge in practical ways. It was one thing to imagine all the experiments you could run with a powerful new centrifuge. It was something else to be able to design, build, or modify the centrifuge to enable those experiments. Hood knew he needed both types of people—scientists and engineers.

As at many universities, each Caltech researcher was like a governor of his own state, able to set his own research agenda. Hood wanted that freedom for wherever it might lead. He also wanted the ability to brush off collaborations he perceived as a burden or a waste of time. At the NIH, because he had rare skills in protein chemistry and he had special equipment, others often nagged him for favors to advance their own experiments. Doing favors for fellow scientists was OK, and part of the culture of science, but there had to be limits. Hood believed there was such a thing as too many teams and too much collaboration.

The Caltech biology division had a long history of support for far-reaching visions and independent thinking. One of the division's influential early figures was Beadle, a Nebraska farm boy who fostered a culture of independence, fairness, and methodical hard work. Beadle could also be bold. Most biologists would gain expertise in one specific area—in genetics that might mean corn, fruit flies, or molds. The questions they could ask were limited to those organisms. Beadle resisted developing tunnel vision. It was often true, then and now, that when scientists got comfortable with a particular hammer, every problem looked like a nail. Not to Beadle. He focused on big questions in genetics. Then he would do the hard work of switching his expertise to whichever model organism (flies, worms, etc.) was most amenable to answering the question. Whatever it took to solve the big problem, he would do.

It was a risky proposition: getting a working understanding of multiple disciplines took the combination of an exceptional intellect, determination, and unwavering self-confidence. Anyone who ventured outside his or her disciplinary comfort zone—and into the lion's den of specialized experts—ran the risk of being written off as a lightweight.

This gamble paid off for Beadle. It helped him find links between specific gene mutations and defined biological functions. His work gave rise to "mutational analysis." This approach quickly became used to analyze "such complex processes as memory, learning, vision, smell, and even the way in which the shape of an organism is patterned during embryonic and fetal development," wrote Paul Berg and Maxine Singer in their biography of Beadle. "Other biologists eventually took note of Beadle's pragmatic approach, and to this day follow his lead in finding the right tool—the right organism—to solve a particular biological question."

This was a place where Hood could thrive.

Hood joined the Caltech faculty on July 1, 1970. It was a few months after Caltech professor Max Delbrück, then in his midsixties, won the Nobel for discoveries of the structure of viruses and how they replicate. That year, Caltech mourned the death of Professor Alfred Sturtevant, a geneticist who had worked with fruit-fly legend Thomas Hunt Morgan and, fifty years earlier, had helped establish the Caltech biology department. A year after Hood arrived, Caltech hired two more biologists from the younger generation. The department was entering a transition phase.

Hood's hiring was noted in the Caltech biology annual report. His photo shows him sporting a coat and tie, short hair with modest sideburns, and the usual bright eyes. The department, full of scientists who never treated patients, played up Hood's medical degree. Hood was charged with strengthening biochemistry and advising premed students. It was up to him to figure out how. "No one ever came to offer me a single word of advice," Hood said. "They just assumed you'd take off, write your grants, get funding, and do your thing. And I did."

It was a fortuitous time to embark on a research career. The NIH budget, the lifeblood for biomedical research, almost tripled during the 1960s. Congress tripled it again in the next decade. The pool of biologists seeking federal research grants was relatively small, making for limited competition. When Hood was starting out, the average age of a first-time NIH grant winner was thirty-five. By 2004, it was forty-two, as federal funding flattened out and a larger pool of scientists had to compete for relatively small pieces of the pie. Many young scientists, especially those with families, found it hard to get started.

At thirty-two, Hood was ready to chart his own course in science. He was married, had two small children, and had landed a plum faculty position at one of the world's great research centers. He didn't realize it then, but the timing of his arrival in greater Los Angeles was also fortunate. Still accustomed to the starvation wages of student life, the Hoods lived frugally. Valerie stayed home with the kids. With help from relatives, they scraped together a down payment for a two-story, three-bedroom Spanish Colonial home in Altadena, a fifteen-minute drive from Caltech. The neighborhood had wide streets with palm trees. It was a middle-class neighborhood where kids could safely run around. The sticker price was $45,000 (about $282,000 in 2016 dollars). It was a stretch.

At times, Hood looked more like a graduate student than a professor. He didn't try to look older and more distinguished, wearing the usual academic tweed jacket. One family photo that year shows him sporting a crew cut, a casual white collared shirt, and gray slacks. He is standing next to his father. Thomas, then in his late fifties, stood with military upright posture, unsmiling. The rest of the family looked happy and relaxed.

From the start, Hood focused on research, but he found more time than usual for the often-neglected task of teaching. He taught a heavier-than-average course load, with classes for graduates and undergraduates in molecular immunology, biochemistry, and molecular evolution. Teaching, thankless and time-consuming as it could be, was critical for Hood's professional growth. It exposed him to a variety of topics in biology and forced him to read widely from the scientific literature. This helped him spot opportunities. Preparing lessons demanded that he clarify his own thoughts. Explaining subjects in clear language for people with varying degrees of sophistication became one of his most valuable skills. Throughout his career, Hood used his rhetorical gifts, honed during his years of teaching, to win over fellow scientists, donors, and the media. "[Teaching] really pushed me to form a worldview and get interested in a lot of different things," Hood said. "You don't fix knowledge [in your mind] by taking courses, but you do fix knowledge by teaching."

Many textbooks at the time were stuffed with jargon. Professors lectured. Hood wanted to make learning more fun. He became friends with fellow professor Bill Wood, another young biologist enthusiastic about breaking the old molds of teaching. These two aspired to create a more engaging kind of textbook. They spent summers working on it at a retreat in Aspen, Colorado.

Students fondly recalled a game Hood and Wood created called Krebs Cycle Poker. It took deadly dry material on a familiar biological process and made it into something more interesting. As in poker, students learned that things in biology aren't always as they appear. The game required students to think analytically, and quantitatively, taking variables into account. One of the lessons, Wood recalled, was "do not bet on your apparent straight unless you are sure about all the intermediates."

Biochemistry: A Problems Approach, by Bill Wood, John Wilson, Bob Benbow, and Lee Hood, was first published in 1974. It became a hit with undergraduate and graduate students. Students were encouraged to think and ask questions. Then they would start trying things in the lab. As Wood said later, they drew inspiration from an old Chinese proverb.

I hear, and I forget.
I read, and I remember.
I do, and I understand.

Hood went on to cowrite textbooks on molecular biology and, with his friend Irv Weissman, on immunology. Many who had been young biologists in the 1970s remembered the books decades later and sent thank-you notes to the authors for inspiring them. One typical comment was from Richard Gayle, who did his undergraduate studies at Caltech in those years:

Biochemistry textbooks in the '70s were written in an encyclopedic, reference format, with few figures, sounding like a hectoring, know-it-all professor from behind a lectern in front of five hundred students. *Biochemistry: A Problems Approach* is formatted quite differently, like a cheat sheet created by a close friend who was part of your study group.

The genius of the format was the easy-to-digest scientific concepts, supported by a problem/answer section using real-world examples that feeds back into the concepts. Just as a real scientist acts when doing research.

It took a while for Hood's research career to gain momentum. He didn't have a single blockbuster discovery propel him to fame, as sometimes happens with scientists in their thirties. He couldn't help but feel a twinge of envy when his friend Bill Wood was elected to the National Academy of Sciences in 1972, at the age of thirty-four. Hood wouldn't earn that distinction for another decade.

During the early to mid-1970s, Hood's lab was an average-sized operation. There were two sides of the lab. The technology group was modest in size and scope. The larger immunology side of the lab focused mainly on components of antibodies—the Y-shaped proteins that play critical roles in the immune system. For Hood's first five years on the Caltech faculty, his research team had relatively modest output, cranking out about a half-dozen peer-reviewed publications a year on average. His resources, and output, were about on par with that of his fellow Caltech faculty member Eric Davidson.

Both Davidson and Hood were bright and confident Young Turks—intellectual pals, sparring partners. The two came of age in an era when biology was becoming more powerful, and more reductionist. That meant focusing on single genes in isolation. Many biologists were working on such narrow problems that it was difficult to see where the answers would lead, other than to more esoteric questions. It was reminiscent of an earlier generation of theoretical physicists drilling into the depths of quantum mechanics. Biologists were increasingly splitting themselves off into specialized fiefdoms that had little in common with each other. Hood and Davidson liked the opposite approach. They wanted to look at the trees without losing sight of the forest.

They shared a desire to understand biology in its broader glory: by studying DNA, RNA, proteins, cells, organs, and someday, whole organisms. Since they had complementary skills, they could help each other. "Neither of us was interested in single genes," Davidson said in a 2013 interview (he died in 2015). "We were interested in whole regulatory systems, whole transcription systems. We were

Caltech's premature systems biologists. We didn't call it systems biology back then, but it's what we did."

They were an odd couple in many ways. Hood was skilled in protein chemistry, while Davidson's expertise was in nucleic acids—DNA and RNA. Hood thought most about questions in immunology; Davidson was a developmental biologist. Davidson's research focused on the genetics of an accessible model organism, the sea urchin, while Hood wanted to cut to the chase in higher mammals. Davidson had long, bushy hair, sideburns, and an unkempt look; Hood was clean-cut, wearing plaid shirts and shorts that made him look like a Boy Scout. Davidson could be profane; Hood's vocabulary was PG. Davidson was a micromanager with such a strong personality that his students would sometimes, creepily, mimic his mannerisms and appearance. Hood was hands-off, the opposite of a micromanager. Davidson, a perfectionist, could be hypercritical of students; Hood was eternally optimistic and upbeat, uninterested in details. The two became lifelong friends and allies.

Hood continued to collaborate with Dreyer, his former adviser—and one-up him if he could.

In 1971, most biological experiments involved tedious manual labor. Many measurements were recorded in lab notebooks with pen and paper. Hood imagined a day when he could more efficiently analyze growing volumes of data spit out from biochemical instruments. The Swedish biochemist Pehr Edman and his collaborator Geoffrey Begg had paved the way for one such instrument: an automated method for sequencing proteins. It used a novel "spinning cup" device that forced the proteins into a thin film inside a cylinder. The force of the spinning made the proteins easily accessible to special chemicals, or reagents. Reacting with the proteins, the chemical reagents would be used in a series of reactions to consistently cleave off an amino acid residue found at a specific spot called the "amino terminus," sometimes known as the N-terminus, found on the end of a protein. That chopping reaction would leave the protein intact and expose a new part of the protein for sequencing. The Edman-Begg degradation process could be repeated until it had been done enough times to reveal the full order of amino acids—the sequence—that made up the full protein. A series of pumps and valves controlled the flow of reagents, while delicately leaving the amino acid residues intact. The first protein sequencing machine using this method was commercialized and made available by Beckman Instruments in the early 1970s. Still, once an investigator had the protein derivatives, analyzing them was a time-consuming chore.

Hood, like Dreyer, wanted to speed things up. Hood had a small lab—just a couple of graduate students and postdocs—with support from the NIH. He sought to build off the Edman-Begg machine, by improving an automated gas chromatograph that could more quickly analyze amino acids, the building blocks that make peptide molecules and larger proteins, such as antibodies. Hood's lab built a prototype that recorded data on magnetic tape cassettes once a day. Data was transferred to a time-sharing computer terminal. Further analysis of the amino acids was done by another Beckman instrument.

Hood remembered one critical piece of advice from Dreyer: if he wanted to shake things up in biology, he should develop a new technology. Hood resolved to spend about half his time on biology research and the other half on developing new technology. Dreyer, following his own advice, also was trying to push the edge with protein sequencing. He served as a consultant to Beckman Instruments, as it sought to improve on its Edman-Begg "spinning cup" device.

Dreyer's ability to think visually compensated for the dyslexia that made him, by his own admission, terrible at spelling, arithmetic, and memorization. "When I'm inventing an instrument or whatever, I see it in my head and I rotate it and try it out and move the gears. If it doesn't work, I rebuild it in my head," he said in a Caltech oral history interview in 1999. He also had an ego. He compared his ability to think in creative visual images to that of Albert Einstein and Richard Feynman. Later in life, Dreyer attributed his inventiveness to this unwavering belief in his own abilities. "Call it arrogance or whatever—it was just plain experience," Dreyer said. "I see it in 3-D Technicolor." This gift, Dreyer said, also made it hard for other biologists to understand what he was seeing, or what he was talking about.

Although Hood recognized his former adviser's gift, Hood also saw a thinker, not a doer. Someone was needed to mobilize the people and resources to do something with those ideas, to prove they were correct. "Dreyer couldn't get anything done," Hood said.

That wasn't entirely true. Dreyer's name alone appeared on a patent issued on December 27, 1977, with the enticing title: "Peptide or Protein Sequencing Method and Apparatus." This was an important invention, and it was Dreyer's. But taking an invention along the rocky road of development to create a viable product—a genuine innovation people could use—was not Dreyer's strong suit. Dreyer claimed he'd made the invention on his own time, as a consultant, and therefore Caltech had no rights. It was his invention. The original license for further development was provided to Durrum Instrument Corp. The company went nowhere.

Hood smelled opportunity. He was convinced he could improve on the apparatus and translate it into a real thing, not just a clunky prototype with tubes and wires hanging off it. He thought he could get things done when Dreyer couldn't. That attitude later led to much friction between the two.

Seeing what Dreyer was doing, Hood wanted to develop a faster, more sensitive, automated sequencer. The original Edman-Begg method had a major limitation. It required large volumes of samples. Biologists are often limited by working on tiny samples from mice.

Researchers needed to get the maximum amount of information out of the samples they had. "I was very excited about the protein sequencer," Hood said, "because I saw if we could sequence very small quantities of proteins, then we could look at whole classes of proteins that hadn't been characterized, and we could open up whole new areas of exploration."

The work was slow, the payoff not immediate. The earliest prototypes were full of bugs, as is often the case. Some biologists dismissed what they saw as mere engineering, and they certainly didn't see the machines opening up new frontiers for their research. There was more than a little snobbery at play. Technology was thought of as "crank turning," best left to industry hacks. The leading men of biology believed their work on the inner mysteries of genes and cells was a cut above. The stereotypical tension between scientists and engineers was still fodder for comedy on *The Big Bang Theory*, a hit TV show, more than forty years later. As character Sheldon Cooper, a physicist, put it, "Engineering: where the noble semiskilled laborers execute the vision of those who think and dream."

As far back as 1973, Sinsheimer, the biology chairman, urged Hood to cut his technology development, to devote himself fully to biology. "His comments didn't have to do with the quality of my work; [they] had to do with my taste," Hood said. The young biologist resisted. He was convinced that his true competitive edge would be in developing a tool that would give him new kinds of data. He knew that new technologies would enable scientists to ask and answer all kinds of new questions, driving biology forward. And who better than biologists knew what technology biologists needed? Technology shouldn't just be left to people who didn't really know how to use it to advance biology, he argued. Most colleagues were unmoved.

The philosophical divide between Hood and many colleagues would only grow.

Defying the boss didn't cost Hood a bit. A couple months after Sinsheimer's warning about technology, Hood was promoted from assistant to associate professor with tenure. He was just thirty-five, with only three years of faculty experience. Freedom and job security were his.

Hood decided tradition and hierarchy were for other people. He perceived himself as special, entitled. If he wanted a piece of equipment and couldn't get a peer-reviewed grant, he'd just raise the money himself and buy it. If an instrument didn't exist, he'd ignore the naysayers, raise the money himself, and pull together a team of graduate students and postdocs to invent it. If a grant was supposed to be used for Project X, but he also wanted to do an unfunded Project Y, he'd use the money to do both. While entrepreneurs and celebrities may become accustomed to preferential treatment, it wasn't common among biologists. Caltech, however, tolerated the eccentricities and excesses of its star faculty. The brilliant chemist Linus Pauling, for example, left many irritated colleagues in his wake when he departed Caltech in 1962. Pauling was widely admired for his discoveries, but many colleagues were less fond of his crusading politics against nuclear proliferation.

As a young professor, Hood started to gain a sense for how power is accumulated and wielded in science. He joined groups of peer reviewers who evaluated grant proposals from scientists all over the country. In 1972, he joined a study section in human biology for the National Science Foundation. A year later, he took on more grant-reviewing work for the American Cancer Society's California branch. Neither group had the money or clout of the NIH. The work was time-consuming, behind-the-scenes, and dull. Still, it was an exceptional opportunity for learning how money gets divvied up in science. Whether he intended it or not, serving on study sections gave Hood a sense of where biology was heading. He got intelligence on what competitors were working on before they published their results. He gained a sense for why some grants passed peer review and why others failed—insights he could use in later years to give his own proposals an edge. Importantly, it helped him build relationships with leading figures. By the end of the decade, Hood's powerful network of friends ushered him into the inner sanctum. He was invited to join the visiting committee of biology advisers to MIT.

In 1974, Hood won his first significant award at Caltech—the Camille Dreyfus Teacher-Scholar Award. It was for "recognition of his research achievements, his innovation in teaching, and his wide-ranging contributions to student life," according to the biology department annual report. A year later, "associate" was dropped from Hood's title, making him a full-fledged professor.

While Hood was striving to advance, biology was getting hot. The 1970s were a time of thrilling possibilities. Scientists were raving about the new gene-splicing techniques, also known as recombinant DNA, pioneered by Stanley N. Cohen and Paul Berg at Stanford and Herbert Boyer at UC-San Francisco. The first monoclonal antibodies—the so-called magic bullets that many scientists envisioned would seek and destroy cancer cells while sparing healthy tissue— were created by Georges Köhler and César Milstein in the United Kingdom. It was now within the realm of imagination that scientists could take a human gene with an important function (such as insulin production), insert it into the nucleus of a host cell from another mammal or from bacteria, and create industrial quantities of the protein that patients could take as a drug. By the early 1980s, scientists would achieve exactly that. Diabetes patients could take injections of human-engineered insulin, instead of insulin extracted from pigs. Not only were supplies easier to obtain, but the body better accepted the human hormone. Genetically engineered copies of human insulin were less likely to provoke the immune system to attack the drug—a regular occurrence with swine-derived insulin.

A new industry, biotechnology, was born. It both thrilled and scared the public, which imagined genetically engineered creatures akin to Frankenstein's monster. "God only knows what's going to crawl out of the laboratory!" the mayor of Cambridge, Massachusetts—home of Harvard University and MIT—once famously said. For a while in the 1970s, the city placed a moratorium on recombinant DNA experiments because of ethical concerns. One of the early biotech pioneers, Harvard's Tom Maniatis, joined Hood on the Caltech faculty in 1977 to continue his research.

As an immunologist and protein chemist, Hood was nowhere near the leading edge of the genetic engineering movement. But he was paying attention. Everything was connected, after all: the genetic code carried the instructions for making proteins. Knowing how to connect those dots was essential for a fully integrated, big-picture view of biology. Hood's discipline was notoriously difficult, intimidating even to biology insiders. It posed a stiff challenge for Hood—staying on top of immunology while simultaneously learning about the DNA and RNA revolution.

Hood knew he couldn't do it all alone. If he was going to advance the field of biology, he would need to deputize other people with complementary skills, especially in engineering. These were the kind of chemically and mechanically inclined people who could tear apart machines and build them into something better. Hood was not that person. Some colleagues would joke in later years that he could barely change the oil in his car.

To achieve these goals, Hood needed to turn on the charm. Like many young faculty members, he had everything to prove. He craved the spotlight and the respect of his peers. When a scientific or business group invited him to speak, no matter how obscure or distant the group, he would get on a plane and go. When journalists called, even from the most obscure publications, including the hometown *Pasadena Star-News*, he gabbed when most colleagues wouldn't bother. Hood's former college roommate, Eric Adelberger, said he saw his friend become increasingly obsessed with proving himself over and over.

Each time Hood traveled somewhere, he did what major scientists do. He tooted his own horn about the work going on in his lab. He widened his network of scientific peers who could help him win grants, pass peer reviews in top journals, and fill him in on the latest scientific trends. He caught up on gossip. But Hood went a step further. Often, wherever he was, he asked to meet privately for an hour with graduate students and postdocs after his lecture. He looked for people who were not just bright—they all were—but for those who shared his enthusiasm. He sharpened his pitch throughout the 1970s, letting students know they had a chance to be part of something big. "I'd tell them about Caltech, tell them about the environment my lab had, and the things I was interested in and not yet talking about," Hood said. "We had great conversations."

During the mid-1970s, Hood's lab started to grow. While many labs have only a handful of students, postdocs, and technicians, Hood enlisted as many as twenty. One of the engineering types Hood sought for his lab was just down the hall.

A soft-spoken young man from Oklahoma, Mike Hunkapiller had gotten his PhD in chemical biology at Caltech in 1974 under Professor Jack Richards. Hunkapiller hung around Caltech for a couple of years as a postdoc, auditing a few courses, trying to figure out what to do with his life. He took a course from Hood in immunology. Hood made it sound fascinating. "He's incredibly energetic and enthusiastic about the science," Hunkapiller said. As a graduate student, Hunkapiller had focused on how enzymes bind with substrate molecules. He became intrigued with similar concepts from immunology—such as how an immune system antibody binds specifically with a foreign substance, known as an antigen. And Hood became intrigued by Hunkapiller. Here was someone with a rare blend of chemistry and engineering expertise: a student who could tackle key instrumentation challenges that he couldn't.

Hunkapiller joined Hood's lab in 1976 and almost immediately became a franchise player. They were a good match. Hunkapiller saw Hood as a visionary who knew what to do in biology but needed help in chemistry, technology, and engineering. Hood had an exceptional ability to zero in on the big problems, define them clearly, generate excitement among financiers, and recruit the people needed to execute the vision. The more reserved Hunkapiller focused on the technical and operational aspects of building the machines—and happily stayed behind the scenes. He was that rare breed of deputy: capable, trustworthy, and not itching for a faculty job somewhere else.

Not long after joining the Hood lab, Hunkapiller got a job offer few starving postdocs could turn down. The Oklahoma Medical Research Foundation, an institute in his home state, offered him a faculty job. It wasn't the Ivy League, but it was a well-established research center and a place to get started as an independent scientist. The offer sent a shiver through Hood. He took Hunkapiller and his wife, Beth, out to dinner. If Hunkapiller stayed at Caltech, Hood told him, together they would be able to do great things, crack open whole new fields. The Oklahoma institute lacked resources and was isolated. The action was at Caltech. It wasn't just bluster. Hood offered a competitive salary. Excited by the new opportunity and the appreciation of the boss, Hunkapiller and his wife chose to stay. They bought their first house in Southern California.

"Lee kind of made one of these offers you couldn't refuse," Hunkapiller said. "He had a vision that there were problems yet in biology you couldn't really study because the tools for doing the analysis weren't up to speed." Even with a somewhat ambiguous position in the academic hierarchy, Hunkapiller stayed seven more years. Long after it was obvious that Hunkapiller had outgrown the position and needed a new challenge, he found it hard to leave. "Mike Hunkapiller was a dominant force," Hood said.

Hood's work habits started to become the stuff of legend as the lab scaled up. He'd often do his scientific reading until 11:00 p.m. or midnight, then sleep four to six hours. If he flew to, say, Tokyo to give a talk and arrived home at 2:00 a.m., he'd sometimes go straight to the lab, showing hardly a trace of jet lag, eager to find out what happened while he was gone, former students say.

On more typical days, he would pop out of bed and go for a run, usually alone, three to six miles. Hood liked running on the dirt track on campus. Partly it was because he had achy knees from his football days. He had also seen many aging, arthritic knees in medical school. Still seeing himself as a football player, Hood continued his fitness routine every day, at home or on the road. Hood would do between 120 and 150 push-ups in a row—a grueling regimen even for a US Marine. He would do a hundred sit-ups, plus other stretching exercises, handgrips, and wall sits—calisthenics exercises he could do without wasting time getting to a gym. Partly, exercise helped him stay ready for his next mountain adventure. It helped him maintain the stamina to get through long days. As his schedule became cluttered with distractions and meetings, he became convinced that he did his best thinking during those solitary morning workouts.

After his workouts, he'd come home to shower, eat a light breakfast, see Valerie and the kids, and usually arrive in the office by 8:00 a.m. Other times, he'd go straight to work, changing out of his sweaty running clothes there. Early in the morning, he'd chitchat with grad students and postdocs who had been working all night and were ready for some sleep.

Hood also kept his finger on the pulse of the lab by inviting his people out to lunch. Or, rather, eating lunch with them. He rarely picked up the tab for his impoverished students, and even lab retreats were turned into potlucks. One favorite spot for cheap eats was Burger Continental. Hood was never a fussy eater. Mitchell Kronenberg, who came to Hood's lab as a graduate student in 1974 and stayed eleven years—an eternity in academic science—looked back fondly on those conversations. "He's actually a pretty good listener. He doesn't always have to show he's the smartest person in the room or be someone who has to have the last word," Kronenberg said.

Kronenberg, who went on to lead a private research institute in San Diego, credits Hood with inspiring him to pursue a career in immunology. "He respects individuality and lets people follow their own path." Hood was serious but could still take a joke. Once, during a skit at a lab retreat, Kronenberg played the role of Hood, dressing in a plaid shirt and shorts to mock his adviser's fashion sense. Everyone laughed, including Hood.

Aside from lunch conversations, Hood developed a reputation for being a hit-or-miss graduate adviser. While some faculty members saw themselves as mentors of grad students, or as hands-on directors of experiments, Hood showed little interest in details or in nurturing. It was the way Dreyer had managed his own graduate students, including Hood. Those who delivered promising results had his ear; those who didn't became invisible. If a student struggled with the technical aspects of an experiment, that was tough luck. He could always reread the literature until he figured it out. Hood didn't want to hear bad news or complaints.

Independent as he was, Hood didn't seem to realize that part of his job was helping students become independent scientists. Hood was a prolific and clear writer, but his former students say he didn't help them learn how to write a winning grant proposal or a paper that could pass scrutiny at top scientific journals. The sink-or-swim environment that Hood created tended to benefit people who arrived self-confident, self-motivated, and self-directed—like himself.

Hood was often insensitive to the struggles of anyone who didn't fit that profile. Tim Hunkapiller, Mike's younger brother, joined Hood's lab in 1978 and witnessed the fallout. "He had people who completely crashed and burned because they were in the wrong lab," the younger Hunkapiller said. "They hated Lee. I'm not sure he ever, to this day, understood it."

Years later, when accepting the Kyoto Prize, Hood reflected on his management style:

> A mentor should provide a rich scientific/intellectual environment, adequate technical and financial resources, and an interesting problem. ... The mentor must be freely available, but in the end, young scientists learn science by working through the challenges themselves. Students must be given the freedom and time to figure out how to solve their particular problem. Once the data are gathered, the mentor again plays an important role in modeling the way in which data is analyzed and, ultimately, the way papers are written. The better the student, the less explicit input is required from the mentor.

Hood could easily have been talking about his parenting style, too. At home, Hood was kind to his children but distant and obsessed with work, much like his own father. As a small boy in grade school, Eran Hood recalls waking up early in the morning to read the newspaper sports pages. It was one time when he knew he had a chance to get close to his dad. Often, Hood would sit at the kitchen table, dictating scientific memos and manuscripts into a microcassette recorder.

Unlike his father, Hood didn't drink alcohol at home. He wasn't reckless or abusive. Neither of Hood's children recalls serious arguments between their parents. Still, Hood's grueling travel schedule clashed with family life. He just wasn't around. Discipline, to the extent it was meted out at all, tended to be handled by Valerie.

When he was home, the children recall, their dad would tell them bedtime stories about a fictional character he called "Harry Golden" who was always getting into sticky situations. Although the children didn't know it then, Hood, an avid reader of science fiction, often would adapt those stories. Harry Golden would sometimes travel in a time machine, and the kids would have to find a way out. Hood was trying to get his kids to think like scientists, by asking them to form a hypothesis, ask questions, and test their imaginations.

Hood and Valerie passed down other values instilled in them while growing up. Education was paramount. Kids were expected to behave. Independence was critical. Children as young as six or seven were expected to run around in the neighborhood and play, unsupervised, and then come home on their own when it got dark and was time for dinner. Their upper-middle class suburb of Los Angeles in the 1970s seemed safe enough.

The Hood household was chaotic, with a steady stream of guests. An extra bedroom was rented out to a rotating cast of characters, including a visiting student from Kenya, a theology student, a family cousin who attended nursing school nearby, and various Caltech postdocs. These people had keys to the house and shared space with the family, Eran Hood said.

"Compared to other kids growing up, I had a tremendous amount of latitude to do whatever I wanted," Eran said. "I never in my life had a curfew. They never told me what to do. It was very much about independence. As long as you were doing well, there were almost no rules."

Valerie worked full-time outside the home when the children were in school. For a while, she was a manager at a sliding glass door company. She taught career assertiveness to other women.

She managed the household—paying the bills, doing the taxes, buying the groceries, cleaning the house, driving the kids to school, attending the parent/teacher conferences. Hood was responsible for cooking dinner one night a week. He cared little for cooking. "Do you want waffles or hamburgers?" Eran recalls his dad saying. "Or that would be the night we'd go out to the Acapulco Mexican restaurant, which he loved." It was a family restaurant, with low prices.

Valerie was Hood's most trusted confidant. Although he avoided discussing work around his kids, it was a prominent topic of conversation after they went to bed. He shared all kinds of personnel dilemmas with her. When he ran into roadblocks or had to deal with a difficult person—as was frequently the case—Valerie often helped him find the solution. Sometimes she would intervene on his behalf. "A lot of people who had a hard time with my dad, or didn't like him, usually liked her. I think that was an asset to him," Eran said.

When Hood would come across as hopelessly insensitive, Valerie set him straight. For instance, he sometimes held lab retreats at a huge, eleven-hundred-acre ranch in northern San Diego County once owned by the famous Hollywood director Frank Capra. Tim Hunkapiller recalls one meal that wasn't potluck-style; Hood and his students ate together at a nearby restaurant. Hood wasn't wealthy then, but he was well-off compared to his students. No matter. At the end of the meal, Hood did the math to split the dinner bill among thirty people. "Valerie gets up from her chair, walks over, and says, 'No, Lee, you get the tab,'" Hunkapiller recalled. He did.

Family, food, fashion, and social graces just weren't top of mind. Hood had work to do.

Chapter 7

The First Big Machine

Hood was obsessed with proving the "two genes, one protein" theory that he'd first encountered in graduate school. This was the idea that certain genes rearrange themselves in infinite combinations, and that's part of what gives rise to the body's vast repertoire of antibodies. To prove it, researchers would need to run experiments evaluating antibody proteins and their underlying DNA. Hood thought the current process was too slow.

An early automated protein sequencer was one of the laboratory workhorses. The Beckman Instruments "sequenator" leaned heavily on Edman and Begg's chemistry work from the 1960s. Improvements were made by the early 1970s that enabled scientists to study small peptides—short chains of amino acids. But the instrument was still limited. It wasn't sensitive enough to parse out certain proteins in trace volumes.

Dreyer had been thinking hard for years about the antibody diversity problem and the necessary tools to work on it. He did some consulting for Beckman, and some tinkering on his own time. On December 27, 1977, Dreyer secured a patent on a new "Peptide or Protein Sequencing Method and Apparatus." The newly envisioned sequencer, with a gas-liquid phase chamber, offered up some tantalizing advantages over the Edman-Begg-based technology that Beckman Instruments was selling at the time. The new tool could potentially handle small sample volumes and pick up low molecular-weight proteins and peptides that would otherwise be washed out in the solvents used by earlier machines. Hood's imagination started firing with possibilities. As he sometimes said, the new process was like snipping off individual letters from a string of pearls representing the alphabet, and then evaluating each pearl to see what letter it was.

While the new protein sequencing prototypes were being tested, the commercially available sequenators of the late 1970s still left plenty to be desired. The analytical process was slow. Early machines couldn't consistently identify all the amino acids, even common ones like tryptophan, cysteine, and serine. Early attempts at automated instruments relied on valves that allowed samples and solvents to pass from one filtering chamber to the next without completely draining between steps, allowing impurities to get into the mix. The valves, which used rubber O-rings as sealants for years, couldn't stand up to the corrosive chemical reagents used by the Edman degradation process. Bits of rubber from the rings would sometimes break off, causing contamination. Trace amounts of samples would sometimes get lost. Given all of the potential technical snafus, scientists needed to start with large volumes of samples, which wasn't always biologically possible. They also needed considerable patience. If they had too small a sample for the crude machine, or if a run got contaminated, they were stuck.

Various labs tried to break down the problem into component parts. None tackled them with the sweeping breadth of Hood's group. Mike Hunkapiller oversaw a small group that tore through the hornets' nest of technical problems. They scrutinized sample preparation. They looked at how proteins were purified and separated from the chemical reagents and solvents. They rethought the mechanical action of using pumps to drive biological samples, solvents, and proteins through an intricate series of vacuums, chambers, and valves. Once the sequenator had completed the Edman degradation, scientists then had to analyze the amino acid residues. They used a method known as high-performance liquid chromatography (HPLC) to separate components by absorbing some and not others. The process could be only as good as its weakest link, so Hood wanted both an improved sequencer and a refined HPLC method.

Bit by bit, Hood and Hunkapiller chipped away at aspects of the problem. While Dreyer had cracked the door open with the invention of the gas-liquid phase protein sequencer, Hood and Hunkapiller set about kicking the door down to make it more broadly useful. Hunkapiller was more of the hands-on leader of a small technical team, while Hood was more like the CEO poking and prodding from arm's length. They tried adding radioactive tags to the protein samples, as other labs had done, to improve the instrument's sensitivity. That idea was shelved as too expensive. Instead, they opted for a nonprotein polymer carrier called Polybrene to attach to the proteins in a sample. Crucially, it would prevent the tiniest amount of precious amino acid residue from getting washed out in the process, and it wouldn't alter the proteins themselves. Hunkapiller's team refined the series of pneumatic vacuums and valves. They improved the way chemical reagents were delivered to mix with proteins.

On one visit to a Berlin research lab at the Max Planck Institute, Hunkapiller saw a crucial missing link. It was a valve that could be shrunk to one-tenth the size they'd been using. The valve had to work in a steady flow of tubes and pumps to send samples through without allowing impurities into the mix. The new valves were said to be leakproof. And because the machine used a different type of spinning cup—which extracted proteins more efficiently—they could get by with smaller volumes of solvents and samples. Hood's lab came up with new vacuum and pressure seals that relied on Teflon and a high-performance plastic known as Kel-F. The new and improved seals replaced the old rubber O-rings, eliminating one source of contamination.

At several steps along the way, Hood and Hunkapiller reported on their progress. One paper in 1978 marked their progress. Published in the journal *Biochemistry*, the paper made a bold declaration in its first sentence: "An important new phase of protein chemistry is underway." As a demonstration project, they sequenced a muscle protein found in minute quantities in sperm whales.

After more than three years of effort, at the end of 1979, they were ready to make a much bigger splash. The improved automated protein sequencer could determine the order of all amino acids that composed all kinds of proteins, from enzymes to hormones to antibody chains. It was at least a hundred times more sensitive than its predecessors. It worked consistently on proteins from multiple species, including proteins thought to be of medical interest. No longer did scientists need milligram quantities of samples to do a respectable protein sequencing run. The machine could determine the order of amino acids in a sample measured in micrograms—one thousandth of a milligram. Maximum information could now be gleaned from a tiny sample.

That was a huge deal for science, because many proteins in blood, urine, saliva, or tissues are found only in trace quantities. The scarcity of many essential biological proteins made them impossible to study. Suddenly, new frontiers were open for exploration.

The excitement was palpable. Hood and Hunkapiller shipped off a manuscript describing their "New Protein Sequenator with Increased Sensitivity" to *Science* magazine on Thanksgiving Day, 1979. The patent application was whipped together and officially filed the day after Christmas. The scientific world heard all about it not just in that coming *Science* article, but also in a blockbuster series of three more papers that Hood and Hunkapiller got published in a single issue of *Science*—the February 1, 1980, edition. It was a magnum opus of technology and biology. Cleverly, Hood and Hunkapiller enlisted some well-regarded collaborators from Yale University and the National Institutes of Health to send them samples. They showed off proteins their machine had been able to sequence from several species: mice and humans—in addition to the sperm whale. The papers described the technical aspects of the machine and showed why it mattered for biology. Their ultimate demonstration was to sequence interferon, a red-hot immune system molecule thought of in those days as a potential treatment for both the common cold and cancer. As Hood and Hunkapiller wrote in *Science*:

> Interferon was discovered in 1957, and, despite its
> potential as both an antiviral and antitumor agent,
> has remained uncharacterized for twenty-two years
> with regard to its primary structure. ... This lack
> of progress arose from the difficulty in purifying
> and analyzing the very small amounts of interferon
> available. In this report, we describe advances
> in automated amino acid sequence analysis that
> permit us to analyze microgram quantities of scarce
> polypeptides [proteins].

Who deserved the credit? Hard feelings emerged among Dreyer, Hood, and Hunkapiller over the relative importance of each man's contributions to the protein sequencer. None of the four blockbuster papers in *Science* named Dreyer as a coauthor. Dreyer's name alone was on the 1977 patent on the gas-liquid phase protein sequenator. By August 1981, the scientists had reached a détente of sorts. All of their names, plus that of technician Rodney Hewick, were listed as authors on a definitive paper describing the technology in the *Journal of Biological Chemistry*. Dreyer was the senior author, signifying his status as the key conceptual driver.

Hood and Hunkapiller, plus Bill Dreyer and Marvin Caruthers of the University of Colorado, were all tapped by venture capitalists to pool their insights on automated instruments in a new company called Applied Biosystems. Over the next couple of years, engineers from Applied Biosystems, some who had cut their teeth on electronics at Hewlett-Packard, visited Caltech regularly to talk with Hood and Hunkapiller about refining product designs. Dreyer, in those crucial early days of the company, was on the sidelines. He suffered from a bout with colon cancer. Years later, Dreyer complained that his contributions had been diminished. Applied Biosystems ran with and improved on "the instrument I invented—just me, not Lee Hood or anyone else," Dreyer said in his 1999 Caltech oral history.

Although Hood didn't create the protein sequencer singlehandedly, his contribution to the new machine made his lab one of the most sought-after in the world. Scientists everywhere started shipping him samples, hoping that the sensitive new prototype protein sequencer could help them solve all kinds of elusive research questions. By the summer of 1981, the Caltech team was focused on miniaturizing the new tool and making it run faster—completing a chemical degradation run in less than an hour. The first commercial version of the protein sequencer—the Model 470A made by Applied Biosystems—was first shipped to customers in August 1982. The tool was an instant hit. It generated $5.9 million in sales its first full year on the market, 1983. That same year, the company sold shares to the public. Sales tripled in 1984. The stock soared. Applied Biosystems poured some of its profits back into Hood's Caltech lab to keep the innovations coming. Applied Biosystems made its mark, living up to CEO Sam Eletr's founding prediction. "We're going to make the picks and shovels for the biotech gold rush," he said.

The tool started arriving on lab benches at the dawn of the biotech era. Scientists imagined all kinds of new experiments they could run. "The machine was like a booster rocket to microbiological research," wrote Bill Dietrich, a science writer at the *Seattle Times*, a decade later. By taking a protein sample and determining its amino acid sequence, researchers could work backward. They could figure out the protein's structure. They could also develop DNA probes to help them clone the underlying gene for the protein. Once researchers had cloned the gene, they could use recombinant DNA techniques to splice the gene into the nucleus of a host cell and coax the cell's machinery to pump out copies. Presto! Instead of extracting trace proteins from animal blood, scientists could make genetically engineered copies and scale them up in industrial-sized vats. Other people in the biotech and pharmaceutical industry could then figure out which proteins had value in the long, difficult journey of drug development.

Many of the proteins now within reach were of little interest to medicine. But some were. Studying the viral proteins involved in inflammation of the brain, or encephalitis—once deemed impossible by Caltech researcher James Strauss—was now possible. Pumping tiny samples through the improved tool, Hood and Strauss determined the sequences of two chains of viral proteins that were fairly large—sixty amino acids long. The discovery was the start of the often-long journey of translating basic science into a practical benefit for patients. In 2009, a vaccine for Japanese encephalitis, based on deep understanding of viral proteins, was approved for sale by the US Food and Drug Administration.

Armed with the powerful tool, and unsure which proteins had medical value, Hood and his team examined many of them. They gathered the protein sequence for erythropoietin (the hormone that stimulates red blood cell production), various interleukin proteins (important in inflammatory diseases), and colony stimulating factors (which stimulate production of infection-fighting white blood cells). Some of those proteins turned out to be useful as drugs.

Hood's approach of combining science and technology was validated. He thought about the impact of the sequencer across science and industry. About where research might go longer term. About the far-reaching impact of his work on society.

Emboldened by the success with the protein sequencer, Hood started thinking more ambitiously about the future of automation. He imagined three other instruments working in harmony: the DNA synthesizer, the protein synthesizer, and the biggest of all, the DNA sequencer. These began to take shape as raw ideas on the drawing board. By synthesizing DNA or protein molecules, scientists could make sufficient quantities for evaluation. The other instruments would sequence DNA or proteins at high speed. They could determine the order of amino acids in any protein, and the order of DNA's base units—adenine, cytosine, guanine, and thymine—known as A, C, G, and T. Empowered with such tools, scientists would be able to work on big, cross-disciplinary problems because DNA and proteins could be sequenced quickly, accurately, and cheaply. At its simplest, this idea would enable scientists to both read and write in biology's fundamental language of proteins and DNA.

By 1979, Hood could start to see where this was headed. His vision for more automated, data-rich biology was attracting powerful people. He was one of the Caltech leaders who persuaded donors—including the Carl F. Braun Trust, the Coca-Cola Company, and Dorothy Chandler of the *Los Angeles Times* family newspaper dynasty—to build a $15 million research lab on campus. Ground was broken on the four-story building at the corner of Wilson Avenue and San Pasqual Street in Pasadena in December 1979. Scientists moved in three years later.

The new Braun Laboratories of Cell Biology and Chemistry was billed as the nerve center for cross-disciplinary work in the powerful new technologies for recombinant DNA, monoclonal antibodies, and research instrumentation. Hood was named to head the institute's cancer research center. The rhetoric was breathless. "We're talking about investigations of the fundamental structure and mechanisms of life itself," Caltech President Marvin "Murph" Goldberger said at the opening ceremony.

Hood learned an unforgettable lesson about the importance of raising private money and the world of business. His most far-sighted initiatives would never fly at the NIH, where peer reviewers often took conservative positions, rejecting any proposal that didn't already have hard data to support it. Private foundations and corporations thought more about long-range strategy. They were more supportive of his dreams. "Federal funding has almost always focused on specific kinds of projects. It isn't focused on creating infrastructure—and it's the infrastructure that's so essential to making the house that really good people can work in," Hood said in a 2013 magazine interview. "That was the point at Caltech with the Braun building. Suddenly we had 200,000 square feet into which we could bring all sorts of new people and things. And without that enabler, we couldn't have made that jump. You don't get federal grants and contracts in general to build new buildings and create new visions."

The time was right for Hood's star to rise. Biology had a grip on the public imagination. The new protein sequencer was in some respects to biology what the semiconductor was to the computer revolution. New tools were opening up a new frontier for biology. For a moment, people could forget about the decline of American industry and the rise of Japanese cars. Popular magazine writers declared that this new industry, biotechnology, was America's to dominate. Biologists were suddenly getting the star treatment, attracting investors who wanted in on the ground floor. A new generation of biologists could now become scientific entrepreneurs. Lee Hood was about to go big.

Chapter 8

Building an Empire

Interferon was the hottest biological substance on earth. Back in the late 1950s, a pair of researchers in London stumbled upon a protein that seemed to "interfere" with live viruses and protect cells from viral infections. As more researchers turned their attention to the mysterious substance, it took on almost mythical status. The protein was like a booster shot to the immune system. Imaginations started running wild in the late 1970s and early 1980s. It was a "magic bullet." It could ward off colds. It could be to viruses what antibiotics were to bacteria. It could even attack cancer. Couldn't it? Interferon made the cover of *Time* magazine in March 1980: "The IF Drug for Cancer." "Interferon was something you sprinkled on stockbrokers to make them give you money," one scientist told author Terry McDermott.

But there was a problem: only trace quantities of interferon could be extracted from the blood of mammals, making it hard to study. The only major supplier in 1980 was in Finland. Given its scarcity, interferon was "perhaps the most expensive substance in the world—one trillionth of a gram costing between $70 and $100," according to *Caltech News*. One scientist amusingly observed that interferon cost about $22 billion a pound.

Until the Hood lab stepped in to seize the opportunity, that is. The improved protein sequencer, a modified version of the tool sold by Beckman Instruments, was the only machine in the world sensitive enough to extract useful information from tiny samples. The Hood lab saw interferon as an ideal molecule to demonstrate the practical medical application of its growing technical prowess. Hood's team raced to determine the amino acid sequences of four members of the interferon protein family. When Hood presented data at one scientific meeting, people leaped out of their chairs to photograph his slides before dashing off for the telephones. Eager collaborators beat a path to the Hood lab, dangling hundreds of thousands of dollars, in what he called "a constant parade and charade." *Caltech News*, a promotional but usually subdued newsletter for faculty and staff, gushed that Lee Hood and Mike Hunkapiller were "cracking the code of a wonder chemical."

With the protein sequence and structure in hand, scientists in other labs could figure out the underlying genetic recipe for interferon. Once they had the gene, biotechnologists could plug it into the nucleus of living host cells. The natural machinery of the host cells, kept alive with a nutrient-rich broth in carefully controlled vats, could then be coaxed to pump out copies of human interferon proteins. Scientists at a biotech company, Biogen, soon declared victory. Human interferon copies could be made by man. The shortage was over. Companies were in a fast competition to drive interferon into clinical drug trials.

Hood watched the scrum from a distance. A scientific frontiersman naturally had to keep moving on to the next thing, especially once settlers arrived. It was a fortunate decision. Interferon, in various forms, was too toxic. It achieved modest victories against viral infections and multiple sclerosis. It never amounted to anything for cancer.

Before 1980, "biotechnology" was just a vague new term, a merger of "biology" and "technology." But now, with more than a little Wall Street hype and entrepreneurial hope, the word took on deeper meaning. Biotech was all about taking ideas, methods, and tools—mostly from elite US universities—to lay the foundation for useful new products. The possibilities were tantalizing. Instead of attacking a limited number of diseases with conventional pills made through chemical synthesis, biotechnologists could use genetic engineering to make whole new types of drugs. These engineered protein drugs could replace crude pharmaceuticals derived from animal blood. The drugs could fix faulty processes at the cellular level in novel ways. For people with all kinds of diseases thought to be untreatable, hope arrived. Investment bankers and entrepreneurs stampeded into the new industry, selling hope of treatments for scourges like cancer, arthritis, and diabetes.

Biotech promised to improve the nation's health, create jobs, and build wealth. The American public was ready for the message. The public had endured a string of humiliations that included Watergate, Vietnam, oil crises, runaway inflation, and the sense that the Japanese were taking over the global economy, beating the United States at its own game in high-profile industries such as automobiles and electronics. Ronald Reagan, winning the White House in a landslide in 1980, promised to undo many regulations and unleash America's entrepreneurial spirit.

The public mood turned more optimistic. Biotech was poised for flight.

Politics, business, and science all churned together, with each advance adding to the excitement. In June 1980, the US Supreme Court voted 5-4 that genetically modified living organisms could be patented. That case, Diamond v. Chakrabarty, gave venture capitalists confidence they could invest in biotech start-ups without fear of the ideas being stolen by copycats. A few months later, Paul Berg, Fred Sanger, and Walter Gilbert shared the Nobel Prize for work on recombinant DNA techniques and manual DNA sequencing. Labs everywhere raced to adopt the new tools. That same month, Genentech had a sensational initial public offering. The stock started at $35 a share, rocketed to $89, and closed at $71.25 on the first day. It hardly mattered that the company was purely speculative, with no moneymaking products at the time. The venture capital firms that bankrolled Genentech, including Kleiner Perkins Caufield & Byers, were now flush. They prowled university halls for more promising ideas.

Then, another political boost: universities that received federal research grants became legally allowed—encouraged, even—to seek patents on their inventions. President Jimmy Carter, just before leaving office, signed the Bayh-Dole Act in December 1980. The idea was to use American university brainpower to stimulate the stagnant economy. This was all about restoring hope in the wake of a dismal period in US history. America's universities were still the envy of the world, and not easily replicated, so policy makers sought to build on that strength. By allowing universities to get into the intellectual-property business, they could strike deals with companies able to further develop research into products. Taxpayer-funded scientists could have private partners in business. Scientists were now empowered to pursue fame and fortune.

Many scientists were appalled. They saw themselves as truth seekers, doing science for science's sake. Business was all about money. Many scientists believed in openly sharing their ideas and their results (of experiments that succeeded, at least). Business demanded secrecy. When students and faculty quit academia for better-paying jobs in companies, they were going to "the dark side," these scientists would say, alluding to the hit movie *Star Wars*.

Hood saw things differently.

Maybe it was because he'd recently turned forty, that middle-aged moment for soul-searching. Whatever the reason, Hood began thinking more deeply about influence and leadership—and the idea that his career might extend beyond science. In 1977, he had been asked to join the search committee for a new Caltech president. He recalled the experience years later:

> It was a very broadening education, because we dealt with the issues of "What is vision?" and "Where is the institute going?" "What should it be all about?" "What kind of person do we want to have?" I think that people who are on search committees that go through that experience ask themselves all those kinds of questions. You ask yourself what your own role is in this whole thing. It was an enormously maturing political experience. I realized how important leadership is, and how important it was to take responsibility.

The committee picked Marvin L. "Murph" Goldberger. He was a theoretical physicist, a visionary scientific leader who could help set priorities for the institute. Goldberger stayed in that job for a decade. Through the committee experience, Hood developed a feel for power dynamics, which could extend his influence beyond his cramped and somewhat dingy lab at Caltech. He liked the feeling.

Hood was becoming recognized as an authority—"internationally known for his research on the immune system," a local newspaper crowed in the late 1970s. He'd been named to an endowed research chair at Caltech—the Ethel Wilson Bowles and Robert Bowles Professor of Biology, to support medically oriented research. And he was becoming more of a public figure, a pied piper of biology.

Hood positioned himself as a helpful interpreter of science for a curious but bewildered public. He honed a message that could resonate in the ivory tower, in the high school classroom, on Wall Street, and on the front page of newspapers. He had ready-made answers, often in the form of easy-to-digest bullet points, prepared for almost every question people would throw at him. He wouldn't spend much time rehearsing his talks, but once he had a sense that a line worked, he would consistently repeat it. Always with an eye on securing financial support for his research, Hood spread the gospel. He spoke in concrete, black-and-white terms. For example, a deeper understanding of immunology could be translated into more effective organ transplants, with less risk of rejection. It could lead to new drugs to tamp down autoimmune disease. It could dial up the immune system's firepower to attack cancer. Knowledge made almost anything possible. He made it all seem thrilling.

Nothing was ever described in a half measure. Progress was "ABSOLUTELY remarkable" or "AN ENORMOUS success." A new development was "INCREDIBLY exciting," or "ABSOLUTELY fascinating," or "transformational," or "revolutionary," or "absolutely superb." He incorporated business buzz phrases, like "paradigm shift," into his vocabulary. He might mention the caveats or nuances in a given scientific discovery, but they wouldn't dampen his excitement. He memorized his points, always speaking without notes—the way his high school debate coach taught him. He'd raise or lower the cadence of his voice to make a point, which almost physically conveyed the joy of discovery. It might have sounded like hype, but Hood could always explain why he thought something was impressive and provide the facts that made him so optimistic. His enthusiasm was infectious. "You won't find a better communicator than Lee Hood," said Pat Gearhart, a Johns Hopkins biochemist. "He can communicate with a grandmother or with a Nobel Laureate. He makes everything he works on sound exciting and crystal clear."

Hood's radiant enthusiasm was unusual. Newcomers to science are taught to be skeptical. They learn to read peer-reviewed articles, listen to seminars, and probe for the soft underbelly in the data, the flawed assumptions, the questions left unanswered. Skepticism serves science well, as only the best theories can withstand scrutiny over time. But scientific skepticism can be taken to an extreme. Biologists can be mean-spirited or cynical as they pick over the details. Hood himself could certainly provide withering commentary on subpar work. If he thought someone or something was wrong, it was "utter nonsense." But he wasn't cynical. He had powerful insecurities, like many other scientists. He fretted about whether his work would make it into the big peer-reviewed journals, or whether his grant proposals would get funding. But he wouldn't let those worries show. He was the irrepressible optimist.

His drive had a powerful effect on people. Enthusiasts were drawn to Hood like moths to a flame. Nitpickers and naysayers weren't welcome. As far as Hood was concerned, especially with the powerful new protein sequencer at his disposal, anything was possible.

As Roger Perlmutter, a postdoc in his lab, said:

> Lee was actually interested in the adventure of
> ideas. If an idea was provocative and interesting,
> Lee wasn't going to sit down and say, "Tell me again,
> how exactly did you purify that? Can you explain
> the details of that protocol?" He wasn't interested
> in those things. ... He was a very different sort of
> person to interact with—a very provocative person.

The investing public was ready to embrace a provocative person with provocative ideas. Biotech offered easy money and impact on the world stage. To Hood, it was thrilling to engage with such a curious, receptive audience. Science didn't have to be about an endless series of esoteric publications gathering dust. If industry could create valuable new drugs, it might plow some profits back into university research. If the fruits of science were made more tangible for the tax-paying public, it might lead to more support for basic research. Futuristic ideas might gain funding, as companies such as Monsanto, Upjohn, and Baxter were sometimes more willing to gamble than supposedly far-sighted academic researchers. Hood—with his sterling scientific credentials, appreciation for technology, optimistic demeanor, and nose for opportunity—was well positioned to ride the biotech wave.

Hood's phone wouldn't stop ringing. Venture capitalists wanted to hear what was next, what else was possible. He loved to talk to the moneymen. He juggled those calls day and night. But Caltech was calling, too: it needed a new chairman of biology. At first, Hood was less interested in that.

There was a leadership void on campus. Robert Sinsheimer, the chairman of Caltech biology who recruited Hood a decade earlier, left in 1977 to become a university chancellor. An interim caretaker didn't want the job. Thirteen outside candidates—including luminaries like Paul Berg at Stanford, J. Michael Bishop at UC San Francisco, Maxine Singer of the National Institutes of Health, and Matthew Meselson at Yale—were approached. None wanted the job.

The search committee then turned to a pair of internal candidates, including Hood. He expressed reluctance. Administration was a chore that distracted from science. But after months of talk, the committee persuaded him. The search committee was led by Eric Davidson, Hood's closest ally on the faculty. Davidson wrote a glowing recommendation to the Caltech president on behalf of the committee. "We believe that [Hood's] national stature as a scientist, his leadership abilities, his organizational skill, his obvious energy and vision are the necessary traits for a chairmanship of our Division," Davidson wrote.

President Goldberger offered the job. Hood held out for a better deal.

In a letter to Caltech's provost, distinguished chemist John D. "Jack" Roberts, Hood said his wife was "strongly opposed" to him taking the biology chairmanship. It would take too much time from family, he said. Hood countered the offer, laying out his terms. He wanted additional money for the biology department, which he could dole out at his own discretion. Part of the money could be used to give raises to three lieutenants who would share the administrative burden. The president agreed to Hood's requests, but the provost pushed back. He argued that biology already had more than its share of "unrestricted" funds.

Hood threw a fit. "I feel as if I was treated in an incredibly shortsighted, unfair, and arbitrary manner. ... I find the administration's hostile response to my requests very difficult to understand," Hood wrote in a February 10, 1980, memo to the provost and the president. "In light of these difficulties, and the fact that I did not want the job in the first place, I am going to take one week to decide whether I will continue to be a candidate." Two days later, President Goldberger passed the memo to Provost Roberts, with a note scrawled in the margin: "Must talk immediately."

Hood got what he wanted and more: permission to skip all faculty senate meetings, which debated everything from social issues like women's rights to campus health insurance plans. "A big waste of time," he said. He accepted the offer.

Goldberger announced Hood as the new chairman of biology on March 24, 1980.

> Dr. Hood is internationally known for his research
> on the immune system. ... His contributions to
> the Caltech community over the years have been
> significant. ... [H]e has organized conferences on
> genetic engineering and behavioral engineering
> at Caltech in an effort to bring science and the
> community together.

The promotion, at age forty-one, added a new dimension to Hood's life. For the first time, Hood wielded real power outside his lab. He had authority over a department with nineteen professors and five emeritus professors. He controlled precious lab space. He oversaw the budget. And he figured the budget should be bigger.

As Provost Roberts recalled in a Caltech oral history:

> Hood would come in—when he did come in; he
> was usually traveling around the world and doing
> great research things—with a set of basically non-
> negotiable demands for appointments or research
> funds for appointments. He was strong for getting
> good people. But his [faculty] appointments were
> very expensive. To my surprise, it turned out that
> getting in biologists was much more expensive than
> the physicists, actually. It took enormous sums of
> money to get biologists set up, but the physicists
> somehow took care of their appointments mostly out
> of their own grant funds.

Hood wasn't the only division chair grabbing for "unrestricted" resources. "I didn't know who was really in charge of the money, or who kept track of what we were spending," Roberts recalled. "I soon found out that I couldn't find out." But Hood was especially emboldened as his profile rose, along with all the talk in the air of biotech revolution. He was the primary fund-raising voice of the ongoing $15 million campaign for the Braun building. The Wall Street money crowd was constantly clamoring for his silver-tongued explanations of the new biology.

The chairman wined and dined with smart, wealthy people. He regularly met with the Caltech board of trustees—a powerful group—and with bankers and industrialists from greater Los Angeles. He schmoozed with elites like Norton Simon, the tycoon who ran Hunt's Foods, and his wife Jennifer Jones, the actress who starred in *Love Is a Many-Splendored Thing*.

Many were beguiled by the smooth-talking yet earnest-sounding Boy Scout of biology. Abe Zarem, a Caltech alumnus who invented a high-speed camera and became a senior executive at Xerox, raved about one of Hood's speeches in a 1982 letter to Goldberger:

> It is hardly possible to imagine a more well put together, informative presentation on immunology. Aside from his extraordinary research record and professional standing, Dr. Hood has to be one of your finer resources in terms of telling the story of research at Caltech.

Hood learned how to take advantage of his growing clout, especially as chairman. And, in several cases, he started to make enemies.

Whenever a faculty job opened, a search committee was appointed to vet candidates. A chairman couldn't pick his guy through executive fiat. But he could influence the outcome by carefully appointing members of search committees. "You could frame the kind of people who would be searched for," Hood said.

Always impatient, and often stubborn, Hood hated to wait for consensus. So he circumvented Caltech's demanding hiring process. He recruited experienced postdocs and let them essentially run their own lab within his lab. The person got considerable autonomy, but without the status or security of a tenure-track job. This caused much consternation and gossip. Some professors wondered whether the recruits were up to snuff. Equally appalling, Hood was challenging academia's sacred hierarchy. At the top of the pyramid was supposed to be a principal investigator who oversaw a motley crew of apprentices: grad students, postdocs, technicians, and senior scientific staff. Having people in unofficial "junior faculty" roles was confusing. Many at Caltech weren't sure how seriously they should take requests from people with vague positions in the pecking order.

Hood also upset people by making tough, long-term decisions for the institute. As a member of Caltech's executive council—the president, the provost, and six division chairs—Hood had a seat at a small, important table. The group met for a full day once a month. Once, Hood joined the majority in voting to boost engineers' salaries above those of scientists. Some fellow scientists considered that an act of betrayal. To Hood, it was simple economics. The job market for engineers was competitive, Hood reasoned, and he didn't want Caltech's best to get poached by industry.

These decisions were just some that created enemies. But Hood also made friends, and they tended to be people with power. Trustees could make Hood's life easier by introducing him to other powerful people, donating to his work, or protecting his turf during budget battles.

To woo them, he needed to learn a bit of tact.

Scientists often value brutal honesty over the social pleasantries that dominate most human interactions. Caltech's Max Delbrück, who influenced the famously cantankerous James Watson, was known for walking out of scientific symposia, telling speakers, "I don't believe a word of it." If he wanted to end a meeting with a subordinate, he wouldn't always bother with common courtesies like "Excuse me; I've got to go." Sometimes he would just pick up the phone and start dialing while the person was still talking. Visitors would awkwardly get up and have to leave without saying good-bye.

Such behavioral rough edges needed polish when Hood was around the philanthropic set. At a minimum, he realized he needed to avoid touchy subjects, either political or personal. Years later, Hood's longtime friend Eric Adelberger lamented that Hood lost some of his authenticity by muzzling some of his strongest views, such as his environmental politics, so as not to offend wealthy donors. Hood himself recalled how he sometimes put his foot in his mouth when in privileged settings. At one dinner party, he asked the "trophy wife" of one of the trustees about the most interesting thing in her life. Her withering response: "There's nothing interesting about my life."

The chairman had more than social graces to worry about. Inside the biology department, he was charged with overseeing several big, pressing issues. The faculty was aging, with five emeritus members—a quarter of the faculty. Existing labs were dingy and cramped, which didn't help with recruiting. The Braun building wasn't yet ready. The faculty, as is the custom in academia, second-guessed many decisions. They were unafraid to share their views.

"Caltech was in a difficult time, a real time of transition," recalled Melvin Simon, one of Hood's faculty recruits, in a Caltech oral history. "The leadership in biology was essentially Lee and Eric Davidson and a few other people, but there weren't very many senior scientists and there wasn't very much leadership—especially in molecular biology."

Hood's most urgent problem was piling up in refrigerators. Scientists from around the world were sending him protein samples, wanting them sequenced. Hood had the only machine, so they had no other choice. Hood couldn't keep up with the demand.

People began complaining that while their particular pet proteins sat around untouched, Hood raced ahead to sequence proteins to advance his own research. The protein sequencer, in other words, gave him an unfair edge—he could leave everyone else in the dust. That's expected in the business world, but in academia, scientists share resources. If Hood insisted on keeping the machine to himself, he would hold back the advancement of science, critics fumed. Something had to be done.

A research center like Caltech couldn't manufacture devices and give them away. This was a job for a company. Hood asked President Goldberger for permission to license the protein sequencer to a company skilled in manufacturing and marketing. Caltech already had a history of productively collaborating with industry, such as Dreyer's work with Beckman Instruments on protein sequencers. Many other Caltech innovations over the years had found broad application in industry, a source of pride. Still, working with industry was touchy. Caltech, like many universities, was trying to carefully calibrate its stance toward what many saw as the "dark side." Choosing the right industrial partner was important. The right company could build a more refined, reliable instrument, spread it widely across science, and make a bundle of money for everyone. The wrong company was liable to put the job on the back burner or botch it altogether.

Who would be lucky enough to get the instrument Hood considered a breakthrough?

The obvious place to make the pitch was Beckman Instruments. Founder Arnold O. Beckman was a force of personality, a man of formidable intellect and business savvy. He was a former Caltech professor—legendary for inventing an electronic pH meter that measured acidity of citrus crops—and a Caltech trustee. He was one of the institute's most important donors. He had funded William Shockley's work in the 1950s, which historians later saw as a key enabling step for the creation of Silicon Valley. The main campus auditorium at Caltech, with more than a thousand seats, was named in his honor. Well into his eighties, he commanded deference among the elites of Caltech. By the time he died in 2004 at the age of 104, Arnold Beckman and his family foundation had given a staggering $400 million to charity, much of it to his alma mater. Before doing anything else, Hood and Mike Hunkapiller spoke to the man himself.

Beckman, then in his late seventies, had retired from day-to-day management of his company, but he retained considerable clout as chairman of the board. Right away, he was enthusiastic. Hood and Hunkapiller were hopeful. But when it came to executing an agreement, the company's middle managers didn't share the old man's enthusiasm.

Beckman's managers had their reasons to be indifferent. The company had long dominated the market for less sensitive protein sequencers. While curious about the newly modified Caltech tool, they suggested that the machine might cannibalize existing sales in a slow-growing market. Hood gave a broad pitch. He shared his dream that a quartet of instruments would revolutionize biology. But the remaining three tools—a DNA synthesizer, a protein synthesizer, and an automated DNA sequencer—were still far from ready for prime time. The Beckman managers didn't see it. "Their reactions revealed a lack of vision on their part," said Mike Hunkapiller, who accompanied Hood on the pitches.

Negotiations stalled. In Hood's view, the company's middle managers had an endless list of demands, they thought Hood's market projections were overly optimistic, and they just didn't want it very much. In Beckman's view, his deputies were "doing a great job." "According to Hood, his side was just angels in trying to deal with them, making every accommodation. Yet, I know they were impatient as hell," recalled Roberts, the former Caltech provost. "I suspect the truth is somewhere in between."

Hood and Hunkapiller drew up a list of other potential partners: DuPont, Becton Dickinson, Johnson & Johnson. More ambivalence. "DuPont nibbled but did not bite," the author Robert Cook-Deegan wrote.

Hood was so confident, he thought he would have to present his case to only one or two companies. He pitched to nineteen. All said no. After one final appeal at Beckman Instruments, a manager said, "Look, Lee, we understand what you're pitching and we're not interested. Don't come back."

Hood was baffled. "Lee's response was always 'They just don't understand,'" said Tim Hunkapiller. "Well, these aren't stupid people. They understand, but they disagree. In Lee's mind, they just don't understand. If only they understood, they'd let him do what he wanted to do. But clearly, in his mind, they don't have enough capacity to fully internalize his greatness or the greatness of what he was doing."

Word about the brush-off from corporate America got around. It reached a budding biotech venture capitalist named Bill Bowes. He asked for a meeting.

Bowes, a man comfortable in the suit-and-tie investment banking world of New York, was instantly amused. Here he was, meeting the great Caltech scientist, the chairman of biology, and the guy was sitting in a scruffy basement lab wearing a plaid shirt, khaki shorts, dark socks, and running shoes. Bowes liked him. The wardrobe reflected a mind so absorbed by biology that little space was left for trivial matters like choosing clothes. It took years for Hood to switch to a more professional outfit of oxford shirts, khakis, and the typical academic tweed jacket. "You have to make him go shopping," said his son Eran. "He'd walk in and say, 'That looks fine, I'll take ten.' He was always a terrible dresser."

Bowes started asking questions. He discovered a rare character who could see far into the future without getting lost in the clouds.

Hood had a practical streak—and a working prototype. The two discussed whether the sensitive new protein sequencer might be the basis for a new company. Bowes especially liked that Hood knew his limitations. He didn't pretend to be a businessman. Hood wanted to provide strategic guidance, but he had no interest in shouldering day-to-day leadership responsibility. "Lee recognized early on that he was not the guy to run the company, which is a bit unusual with founders," Bowes said. "They always tend to think, 'I'm smart enough to found this thing. I'm smart enough to run this thing.' It's hardly ever the case. They don't have the background. … I appreciated not having to fight that battle. It often leads to hard feelings."

Excited, with entrepreneurial fire in the belly, Bowes and Hood went to work crafting a license. Bowes identified a cofounder and CEO for the new company in Sam Eletr. Eletr was a former manager of analytical and medical instruments at Hewlett-Packard. By late 1979, before any venture capital had been invested and when the protein sequencing company was just a raw idea, Eletr quit his job and went to work full-time on just the promise of lucrative founder's shares. Motivated to make the new company a reality, Eletr studied quickly. He rounded up the necessary intellectual property. Many players were involved. Caltech had the rights to a series of modifications Hood and Hunkapiller had made to a Beckman Instruments protein sequencer. Bill Dreyer had an important patent in his own name. The Max Planck Institute in Germany needed to participate. The company would need all the puzzle pieces to have the freedom to engineer a commercially viable protein sequencer. Caltech would license development rights for the protein sequencer to this new company called Applied Biosystems Inc. (later known as ABI).

But Caltech lawyers weren't going to just give away the store. While many prototype instruments are riddled with problems and require serious engineering to become commercially valid, this one was in decent shape and served as a good template for the company to hit the ground running. Negotiations dragged on "interminably," Bowes recalled.

Conflicts of interest were a sticking point. Companies sometimes asked for the right to review, and in some cases censor, academic research they sponsored that might undermine their market standing. Many in academia were suspicious. How would it look, for example, if Hood used his Caltech departmental budget to purchase a bunch of machines from a company he helped start? Harvard faculty had just revolted over a venture-backed start-up called Genetics Institute. Many Harvard scientists argued that allowing faculty to become stockholders and promoters of their own speculative biotech start-ups would surely end up tainting the academy. As Roberts recalled:

> A lot of professors were almost livid about having
> this kind of operation. ... So I finally delivered
> the bad news to Hood that we were just not going
> to want him to be a stockholder in this thing. ...
> I think he understood the ethical problem. ...
> Hood was a bit grudging, but he agreed that he
> probably shouldn't do it.

Hood continued evangelizing for the instrument by giving public talks. At one, Arnold Beckman was in the audience. "This is fascinating. It is just what my company needs," Beckman said afterward. Hood told the man that negotiations with his company had broken down, and Caltech was helping start a new company to commercialize the instruments. Infuriated, Beckman flew to his company's Spinco Division in Palo Alto, California. The managers there "tried to cover their collective rear ends," Mike Hunkapiller said. The managers told Beckman, "'Oh, Lee misled us, because he obviously wanted to start a company and make a lot of money,'" Hood recalled. "It couldn't have been further from the truth." Beckman met with Caltech president Murph Goldberger. How could Caltech create a competitor to seize what Beckman thought was going to be his company's big opportunity? And Hood was one of his favorites on the Caltech faculty.

Beckman pushed for renewed negotiations. It was spring 1981, more than a year after Hood and Hunkapiller first approached Beckman managers. The president was in a tough spot. Goldberger surely didn't want to alienate a big donor, but too many at Caltech had worked long and hard to negotiate favorable terms with Applied Biosystems. They were in final talks. It was unlikely they'd rip up a deal at the eleventh hour. Besides, the Beckman managers weren't offering more enthusiasm or better terms. Caltech offered Beckman a consolation prize. The license to Applied Biosystems would proceed, but Beckman Instruments would get a "nonexclusive" right to commercialize the technology as well. Arnold Beckman seethed over the outcome for some time, but he didn't hold a grudge. A few years later, he donated $50 million for a chemistry and biology facility called the Beckman Institute. Applied Biosystems, brimming with entrepreneurial excitement and focus, went on to dominate the market.

Two lessons stuck with Hood the rest of life. First: new ideas need new organizations. A hungry start-up was better positioned to recruit bright engineers and businesspeople who like tough challenges. Big, diversified companies were unlikely to bring as much focus and urgency to any one mission. Second: deal only with top executives who have the authority to take risks.

Hood had found the right partner in Applied Biosystems, and it made buckets of money for everyone—$5.9 million in sales its first full year on the market, and triple that the next year. Caltech, along with the individual inventors named on the patents licensed to Applied Biosystems, stood to collect a cut from sales of the protein sequencer and subsequent instruments.

The ramifications for the biotech industry were profound. The protein sequencer alone "advanced the whole field, speeded up research by about two orders of magnitude," said Bowes, the venture capitalist. "It took you three weeks to get a result before, and now you could get it in twenty minutes because of his gadget."

The protein sequencer enabled some incredible work. Scientists used it to discover that proteins, called prions, could become infectious from one species to another. Prions are misfolded proteins that can cause brain-wasting conditions such as "Mad Cow" disease, or Creutzfeldt-Jakob disease, in humans. Stanley Prusiner at UC San Francisco was awarded a Nobel Prize for his work on prions.

The machine also enabled the discovery that cancer, a disease in which cells become immortal and proliferate out of control, was related to normal cell growth gone haywire—an unsettling finding. It was 1983, and Hood and Mike Hunkapiller were sequencing platelet-derived growth factor. Along with other researchers, they were able to show that genes for a normal protein—involved in clotting, repairing wounds, and forming blood vessels—could be linked to an oncogene: a gene that causes cancer. The New York Times reported that "for the first time, a close similarity has been discovered between a cancer gene and a natural substance that has a known function in the human body." Hood told a reporter for Science News that the work "makes you wonder all sorts of things, such as 'Are many of the oncogenes growth factors?'" As is often the case, the research raised more questions than it answered. Researchers stayed busy for decades trying to understand the complexity of the genetic variations and environmental stimuli that lead to cancer.

As Hood continued his scientific work and got his next automated machine off the drawing board, he also expanded his business interests. Bowes was helping put together the company that would become biotech's first big success story: Applied Molecules and Genetics, later renamed Amgen. Hood was asked to join the scientific advisory board. He did, along with Marvin Caruthers, a DNA synthesis pioneer; Arno Motulsky, a leader in human genetics; and Hood's friend Irv Weissman, a future stem cell biologist.

At first, Amgen didn't know what to do. The scientific advisers toyed with ideas that sound silly in hindsight. One was to make indigo dye for blue jeans. Another was to make chicken growth hormones. For a while, before Applied Biosystems was up and running, the investors considered in-licensing the protein sequencing technologies to provide a competitive advantage for a fledgling biotech drugmaker. Quickly, the board agreed that everyone would make more money if the protein sequencers were sold by a separate company to researchers around the world. Finally, Amgen CEO George Rathmann and the scientific advisers settled on becoming a drug company with specialized skill in making genetically engineered drugs. Although Hood often liked to credit himself as a cofounder of Amgen, the scientific advisory board met only a handful of times a year, and not always in person. Rathmann, a bear of a man with a warm demeanor, was, by all accounts, the driving force at Amgen. Decades later, when Amgen was established as a huge success, Hood's official biography in his institute's promotional literature said he "founded or co-founded 15 different biotechnology companies" with Amgen at the head of the list. Although it was fair to say Hood was involved in early strategic conversations, it was an exaggeration to imply that he played a central role in that many successful enterprises.

Hood was now attracting attention not only from scientific media but also from popular media. Hood and Mike Hunkapiller got the star treatment in March 1981 when the *New York Times* ran a lengthy science story under the headline, "New Gene Machines Speeding Up the Pace of Biological Revolution." "Someday," the article stated, "there will probably be a library containing all the genetic information needed to create a complete human being."

Years before the Human Genome Project was even a topic of discussion, Hood showed his gift for talking up a tantalizing future without crossing the line into hucksterism. "I think we are only beginning to imagine all the things we can do with these machines," Hood told the *Times*.

Soon, lengthy magazine profiles began to lionize the man with the "gene machine." The *Los Angeles Times* declared in October 1985: "Leroy Hood Built a Better Gene Machine and the World Beat a Path to His Lab." A photo caption carried a breathless overstatement: "Caltech molecular biologist Leroy Hood demonstrates the protein synthesizer, which is to biology what the cyclotron was to high-energy physics." The instruments, at the time, were still quite limited. The cyclotron was an analogy Hood used to explain what his instruments could do for biology. He was photographed with a liquid-squeezing pipette in hand, wearing a white lab coat. Members of the Hood lab shook their heads and laughed. They never saw the boss wear such an outfit when the cameras weren't around. But Hood was always courteous and accommodating to the visiting media. If they wanted him to pose in a white lab coat, he'd do it.

Then Amgen hit the jackpot—and so did Hood. Erythropoietin (pronounced uh-RITH-ruh-PO-eh-tin) is a hormone with an immediately obvious medical benefit: it boosts the production of oxygen-carrying red blood cells. In the 1970s, Eugene Goldwasser, a biochemistry professor at the University of Chicago, began doing the classic, painstaking work necessary to better understand the hormone that went by the nickname EPO (EE-poh). Goldwasser began collaborating with the start-up, and by 1983, an Amgen team led by Fu-Kuen Lin found the gene that provided the code for the precious protein. Amgen got the crucial gene patent, and within a couple years, Goldwasser and Amgen collaborators published the first detailed structural characterization of EPO. This described the protein sequence, the sites at which carbohydrate molecules tend to hang on to it, and the location of disulfide bonds. By the mid-1980s, Amgen was running full speed with the baton. The company learned to manufacture it to the standards of consistency and purity mandated by the FDA, and designed clinical trials to prove it could help patients. Erythropoietin was tremendously powerful. People who were bedridden with anemia, dependent on chronic blood transfusions, could in some cases become well enough to climb stairs and play golf. Later, Tour de France cyclists used it illicitly to boost their endurance. EPO became biotech's first billion-dollar-a-year drug. Amgen became the world's biggest biotech success story. And Hood was in position to become wealthy.

This time, because no Caltech intellectual property was involved, Hood had received founders' stock shares in Amgen. Even though Hood had little to do with the company and left the scientific advisory board in 1989—the same year EPO won approval from the FDA—he shrewdly held on to his shares for more than twenty years. They were an extraordinary windfall, adding up to millions of dollars. Hood was already doing well by the modest standards of those in academic science. As chairman of biology at Caltech, he had been provided a university house within a few minutes' walk from campus. So he and Valerie sold their Altadena home, pocketed a gain of more than $600,000, and bought a vacation home in Santa Barbara. The Amgen shares catapulted Hood's finances to a whole new level.

Wealth didn't change his fashion sense—or his work ethic. Hood kept taking on science, business, and administrative duties. In contrast to some scientists, who take pride in drilling deeply into a narrow set of questions—taking a hypothesis and "beating it to death," one colleague put it—Hood would constantly try to ask an interesting, important question, get the answer as fast as possible, and then move on to the next thing that caught his eye. He didn't want to go all the way down the narrow rabbit hole of elucidating something like prion biology for the next decade. He came up with so many ideas that fit within his conceptual framework of "big science" that it was hard to execute on all of them at once. Some were good ideas; some weren't. But the result was that multiple projects were all spinning at once. Some colleagues, by the early to mid-1980s, thought things had spun out of control.

Hood's lab swelled to eighty and even one hundred members, including interns, at times during the summer. Such a big lab was inconsistent with Caltech's culture of smallness. Most labs had fewer than a dozen people. When Hood began building what looked to some like an almost corporate behemoth in the middle of Caltech's campus, many perceived it as a violation of the campus code. It didn't help when the media started describing Hood's operation as a "gene factory," which called to mind masses of worker bees mindlessly plowing through dull tasks. It clashed with Caltech's most cherished values. "While diminutive scale may be a disadvantage for some institutions, for Caltech, it is at the heart of its being, and perhaps the single most important aspect of its extraordinary global success," the *Times* of London once wrote, ranking Caltech over Harvard and MIT as the world's greatest university.

As Hood's public profile rose, it wasn't unusual for him to be in three or four cities, or two or three countries, in one week. He might jet to London for a day or two, be back in Los Angeles for a day, and then off to New York. Spending so much time alone on airplanes gave him time to think and write. Each day in the new city would be packed with meetings. Hood's apprentices were astonished by the physical stamina his schedule demanded. "Lee and I are criticized continually because we travel so much," Irv Weissman said. "But the facts of life are that nobody really reads the literature, broadly. If you want your ideas to get across, if you want your papers published, if you want to get grants, if you want to get great students, you've got to be on the road at least a third of the time." For Hood, it also didn't hurt to get one more shot of adulation from his admiring crowds.

He already had a reputation as a lackluster mentor, and being on the road constantly didn't help. One year, members of the lab printed up T-shirts with the title "Lee Hood World Tour" and a list of dates and places on the back, as if he were a rock star on tour. One insider, Mitchell Kronenberg, found the cotton T-shirt so amusing that he wore his for years, "until it disintegrated." "We were being affectionate, but snide at the same time," said Tim Hunkapiller, who worked for fourteen years under Hood at Caltech. "He didn't get the joke." Members of the lab, accustomed to being unsupervised, sarcastically called themselves "Hoodlums." Once, someone posted a missing-child poster on a bulletin board with Hood's picture on it. Estimates put him at the lab 30 percent of the time, with 70 percent spent traveling.

Hood didn't always know what was going on in his lab. "He'd say something like 'I know, let's do that experiment' in a group meeting, and we'd say 'OK, but we did that last year and published a paper on it.' He'd just say 'OK, then let's do a different experiment.' It wouldn't even embarrass him," said one former postdoc, Joan Goverman. "He was constantly multitasking and overcommitted. I don't know how much he applied his full intellectual force to any one thing."

On reflection years later, Hood said he did travel too much, but he resisted the notion that he was overextended. "One of the interesting questions I thought about was 'What are the limits of what I can do?'" Hood said. "I never felt I had exceeded my limits. I never felt there were too many things to manage. I still have never felt that. I won't say I always did everything as well as I could have, but from my point of view, in balancing my time and the priorities, I did a pretty good job."

Hood grew even more distant to his team. Like a CEO overseeing a sprawling empire, he started to delegate tasks—especially the unpleasant human ones, like firing people. Hood showed little curiosity about the personal lives of those in his lab. It wasn't his style to pull a student aside and say, "You've got talent; why don't you try this certain thing to develop it." He didn't ask about personal milestones like a marriage, birth of a child, or purchase of a first home. "The essence of academia is essentially a transitory relationship with all the people in your group," Hood said with a shrug. He saw those in his lab as one of three types of people, Tim Hunkapiller recalled: stars, trusted lieutenants, and "bodies." The last group he treated the way the name implies. Hood often couldn't remember their names. "There he is!" or "There she is!" he'd call out enthusiastically when passing someone in the hall he didn't know. More than once, he mistakenly called a young scientist "Dallas" whose name was Austin. Mentoring them wasn't a priority. Their opinions didn't count. As Tim Hunkapiller remembered:

> I was lucky. Lee did have a grudging respect for me and an honest-to-God liking of me. But there were others who really got screwed by Lee, because Lee would go to the person who at the moment gave him the most. If you were part of the third group, and in conflict with one of the stars, Lee was very clear— even if you were right, it didn't make any difference. He played favorites. Explicitly.

The stars were the bright young protégés of biology—people like Roger Perlmutter, who went on to run R&D at Amgen and Merck, and Michael Steinmetz, a future executive at Hoffmann LaRoche and venture capitalist. Rising stars on the technology side of the lab included Mike Hunkapiller, Lloyd Smith, and Ruedi Aebersold. Stars had direct access to the boss. Trusted lieutenants were those Hood couldn't get by without. They could give Hood bad news without fear of the consequences. Hood listened to them and trusted their advice. He also did his best to keep them around. Tim Hunkapiller, for example, arrived as a graduate student in 1978. Normally, students spend three to five years getting a PhD. It took Hunkapiller, technically, fifteen years. He had done the necessary work years earlier. But there was always some important new project he and Hood wanted to complete before he got his degree.

Tim Hunkapiller's knowledge of computational biology was rare; he was early among those thinking about using computers to analyze biological data coming from the automated instruments. Desktop computers were still a luxury in biology labs of the early to mid-1980s. Some were jealous when Hunkapiller got one from the boss and used it to develop an essential skill. In one of Hood's off-the-cuff employment arrangements, Tim Hunkapiller lived in a netherworld as a Hood deputy, neither postdoc nor professor, with status almost like a corporate vice president. "Caltech is a very special place," Hunkapiller said. "I had a lab. I had grants. I had grad students, postdocs working for me, and I wasn't even officially part of Caltech. ... Lee isn't real big on the rules and regs, the Ps and Qs. Caltech isn't, either."

Hood didn't want to lose great people, and he didn't want to waste time. Science was a competitive sport. Being first mattered. Hood didn't have the patience of some biologists, known for methodically designing experiments and then spending years seeing them through in precise detail to their logical conclusions. He wanted experiments to be well designed and carefully controlled, but he also wanted results and wanted them fast. His style was to get some data, publish, and move on quickly to the next big idea. It was how he stayed at the leading edge.

The Hood lab valued urgency like a start-up running out of cash. Michael Steinmetz figured out early on that whatever the boss wanted, he wanted ten minutes ago. As Steinmetz recalled:

When he came to the lab, he'd go around and talk to
everybody. He'd say, "What's new, Michael?" If we
had some substantial results, he'd say, "Let's write
it up, publish it." This would be in the afternoon.
When you came back the next morning, you'd have
a draft manuscript on your table. He dictated on a
tape at night, and then gave it to his secretary. ...
He would have already written the introduction and
discussion [conclusion]. You'd fill in the results.

Rival labs rushed so fast to keep up with the Hood lab juggernaut
that they sometimes made embarrassing errors. Members of the
Hood lab chuckled when they saw competitors submit papers with
spelling mistakes, wrong names, and mislabeled charts. It looked
like people tried to cram two weeks' worth of experiments into just
a couple of days. Hood had gotten inside their heads, rattled them.

Moving at whirlwind speed, with great enthusiasm and disre-
gard for details, stimulated people. But it also created carelessness,
chaos, and confusion. Roger Perlmutter tells a story of trying to meet
Hood for an interview in the fall of 1980. Perlmutter, who trained
as a physician/scientist at conservative places such as Washington
University and Massachusetts General Hospital (MGH), was fascinated by Hood's cavalier, creative style. Trying to figure out his next
move, he asked whether Hood had room for a physician who understood science. Sure, Hood said. Come see me at noon on Saturday at
the corner of Wilson and San Pasqual Street in Pasadena. Perlmutter
flew to Los Angeles, borrowed a friend's truck, unfurled a map, and
drove to the spot. He sat for a while, looking around. He didn't
see anybody. Realizing he was at the edge of the Caltech campus,
Perlmutter wandered into the nearest research building. There was
Hood, in his basement office, writing. Apparently, it hadn't occurred
to Hood to mention the name of the building and the room number.

Hood created an anything-goes lab culture that energized some and bewildered others. The lab was home to disciplined, precise, buttoned-down characters—and equally welcoming to laid-back free spirits. One technician regularly brought a seven-foot boa constrictor to work. She'd ride her bike with the snake, named Snakey, in her backpack, head sticking out. "You didn't have a snake named Snakey in an MGH laboratory," Perlmutter said.

"What was important were your results," said Steinmetz, a buttoned-down type who was in charge of the technician with the snake. Hood seemed to deliberately sow the seeds of chaos. Some thought he was even proud of it. "Lee thinks that anarchy brings out creativity," former postdoc Mitchell Kronenberg once said.

Several faculty recruits were thrilled by the chairman's style. Hood showed Ellen Rothenberg around campus in his standard uniform, the unpretentious plaid shirt and shorts. In his basement office, she noticed scientific papers stacked up on an old couch, alongside sweaty running clothes. This was a place that treated science as a great adventure. It was not the stuffy place Rothenberg saw during her student days at Harvard and MIT. As she recalled:

> The whole East Coast in the 1970s was very focused
> on scientific achievement with angst. We all tried to
> outdo each other by how much we could work, and
> how joyless our lives could be. How passionate, with
> almost suicidal intensity, could you be? This notion
> [at Caltech] that science is actually fun, and that you
> play in between work, was very, very unusual.

If the people in this freewheeling lab culture were a mix, so were the results of the culture. Freedom and autonomy felt empowering to those young scientists with the requisite amounts of motivation and confidence. Hood had an undeniable knack for inspiring bright people to do their best science. He also understood that he couldn't succeed without smart, dependable deputies. He was good at articulating a clear path forward, raising money, recruiting, and creating an environment where people dared to take on big challenges. Like most visionary leaders, he needed managers under him to make sure things got done right.

But trouble was inevitable. A lot of smart, ambitious, egotistical people were packed into a small place. As in every other lab, people argued about how experiments should be done, how papers should be written, and who deserved credit on papers. There were rivalries. Some people hated each other. One leader of a mini-fiefdom within the lab ordered all his subordinates to stop speaking to a member of the lab he disliked. Hood tried to avoid all such emotionally raw conflicts—even when asked to intervene. Sometimes, he allowed conflicts to fester until they became cancerous. Other times, Hood was the source of conflict. Once, a postdoc asked for a raise, and Hood gave his word. When it came time, Hood reversed himself, saying he didn't have the budget. The postdoc, a sole breadwinner with three kids at home, protested, saying a commitment had been made. Hood responded: "Commitments should be flexible."

"That right there is classic Lee," said Joan Goverman. "The definition of commitment is there should be no flexibility! But he was genuinely puzzled by this postdoc's reaction. Commitments, to Lee, should be flexible, because he needs to be able to do what's good for him. It's a selfish perspective—except that Lee doesn't really get it. To Lee, what's good for Lee is also good for you."

Hood had the same attitude at home. As his kids were growing up, Valerie told friends, Hood would agree to share carpool duties, and then try to back out after making another commitment, usually to give a scientific talk.

"Superscientist Balances Home Life and Lab Life." That was the headline in the *Los Angeles Times*, Southern California's most influential newspaper. It was all too perfect. A press conference had been called in Washington, DC, to announce a development in Hood's lab, but Hood insisted the press conference be moved to Pasadena so that he could attend Honors Day at his children's high school.

His daughter, Marqui, was blunt. "He's not the super god that articles make him out to be," she told the newspaper. Hood himself spoke carefully about his roles as husband and father.

> Life is a series of trade-offs. You have to decide what your priorities are. We have not a spectacular, but a quite reasonable, family life. We've raised two kids who are wonderful people. ... But once you get into doing something you really like, it's hard to stop. I don't view my job as a job. It's what I spend most of my time doing.

He didn't escape embarrassing his teenagers. Once at a school picnic, some of Marqui's friends and their parents started a sand volleyball match. She was mortified to see her dad take off his shirt to join the game, showing off his athletic physique. "I think Valerie and I embarrassed her many times," Hood said.

Marqui also went through what she called a "sullen teenage" phase in the last half of high school. She was feeling inadequate, she said, compared with her older brother, Eran. He was a star student, athlete, and outdoorsman, much like his father. Marqui also clashed frequently with her mother. When Hood was home, he told his daughter to be more respectful to Valerie. It didn't help. Hood went with his wife and daughter to a family psychologist. That didn't help. Not understanding what was going on, Hood made the insensitive complaint that it was Marqui who withdrew from the family. She had no reason to worry, he said, because her parents loved her every bit as much as Eran. That didn't help, either. What did help was Marqui leaving for college on the East Coast. She became homesick and called often. "It was like a magical switch," Hood recalled.

To the *Los Angeles Times*, Valerie Hood downplayed any notion of her husband being father of the year. When the reporter asked a softball question—which event had more impact in the Hood household: his new scientific development or the kids' Honors Day?—Valerie didn't take the bait. "It's exciting to share in something that is national news, but you have to realize that there is something exciting happening in Lee's life every day," she told the newspaper. Valerie added: "I think Lee was never here for a parent open house. There are a lot of things like that I've gone to alone. But on the other hand we have had these incredible opportunities."

The writer of the newspaper story was starstruck, like so many others who came into contact with Hood in those years. Delivering talks on the road, Hood left journalists, peers, and younger scientists amazed. Long before PowerPoint and TED talks, Hood became famous for delivering mesmerizing talks that radiated energy and were broad in scope. He'd have not one, but two or three, slide projectors going at once. One set of slides would describe his overarching vision, and others would describe a mountain of data his lab had gathered to support it.

Like the best of stories, his scientific talks had a provocative beginning (often a prediction), followed by a suspenseful middle (the scientific search for answers), and a dramatic resolution at the end (which always provided a springboard for some new scientific quest). Hood looked from left to right, sizing up the audience. Always, he looked them in the eye. He would vary his cadence as he spoke, raising his voice in boyish wonder when making an especially interesting point. His youthful-sounding voice made him come across like an excitable graduate student eager to share his latest results. A surprising point brought an arch to his left eyebrow.

He gestured—not wildly, but more in the measured manner of a technology CEO in command of the room, demonstrating the latest feature of a cool device. Sometimes he'd act as though he had just picked up an ancient dinosaur fossil and was gently holding it up for inspection by the audience, even when he was making a purely theoretical point. Perlmutter joked that Hood was the "hardest working man in show business," a riff on the tag line of soul singer James Brown.

Hood gave the sense that he and his fellow scientists were about to make history. "We will learn more about human development and pathology in the next twenty-five years than we have in the past two thousand," Hood proclaimed in one essay. He always explained the big picture in plain English. He saw the forest and the trees. Nobody ever had to wonder why his experiments were done or what the point was. Each talk had a consistent theme: these were big problems in biology, and solving them was just about the most exciting thing in the world to do.

Decades later, many scientists couldn't remember specifics of the talks, but they remembered the spell he cast. Something deeply emotional—partly manic, partly soulful—was at work. The greatest orators, whether they be politicians, musicians, preachers, or scientists, have an uncanny ability to bond with an audience and draw energy from it. As Robert Plant, the lead singer of the rock band Led Zeppelin, once said: "There is a difference between people who sing and those who take that voice to another, otherworldly place, who create a euphoria within themselves. It's transfiguration." Hood did that. Some in his lab joked that if Hood's science faltered, he could start a religion.

Young scientists outside the Hood lab often walked away more intimidated than inspired. For anybody who dared to compete with his lab's work head-on, Hood's talks seemed to carry an implicit warning: Don't bother. You'll get run over. Weissman said: "I can't tell you how many times [Hood] gave talks at Stanford, or he gave talks at a meeting somewhere, and afterwards, all the young people would be devastated, because he covered in his lecture just about everything they were doing and hoped to do. And he'd do it better."

Dating back to his football days, Hood loved to compete. He loved being first with a discovery, loved seeing it in print. Hood once told graduate student Mark Davis that he gave ambitious talks, stayed physically fit, and slept four to five hours a night all for the same reason: "It adds to my aura of invincibility." It worked. The energy Hood put into everything—his science, fitness, fund-raising, and especially his speaking—astounded those around him. "I've never met anyone with close to that kind of energy," Perlmutter said.

There was just one problem with Hood's talks.

He exaggerated. Often. Hood would say his lab was sequencing proteins at degrees of sensitivity that were premature by a full order of magnitude. Sometimes Hood would use broad ranges that implied exceptional capabilities that didn't exist. Imagine a football coach saying he has a quarterback who can throw the ball fifty to five hundred yards downfield, when in fact, the quarterback can throw the ball fifty yards and anything much farther is impossible. It might become true sometime in the future, and the Hood team was working hard to make it happen, but it hadn't been done. Many of his young charges were appalled at this violation of science's fact-driven culture. It was like Steve Jobs, who famously created a "reality distortion field" around himself to get skeptical employees to suspend their disbelief and work hard on the impossible.

"We'd go to him and say 'Lee, that's not true,'" Goverman recalled. Hood's response: It was no big deal, since it was sure to happen in the next month or so. People wanted to hear big promises, not an endless list of caveats. Hood gave people what they wanted. He certainly wasn't the only entrepreneurial scientist who got carried away in a quest to excite people and raise money. The only option for his team was to try to make the impossible into the possible. "He didn't like to hear bad news," Goverman said.

Then he got some about his family. His father, Thomas Hood, died of a heart attack in March 1985 at age sixty-nine. Thomas was on vacation in Arizona. Lee had been mustering the courage to confront his father and ask why he'd been so cold and distant. Or at least let him know how much it hurt. The conversation never happened. Failing to clear the air before his father's death was "one of the greatest regrets of my life," Hood said.

The father who skipped his son's football games and graduations died just as his son—in his midforties—was becoming a leading figure in science. Life was short, Lee Hood thought. It must be lived fully. His dad didn't live long enough to see his son compete for the biggest prize in science.

Chapter 9

Competing for the Big Prize

How does the immune system work? How does it distinguish between friend and foe—the foreign structures worthy of attack, and the healthy tissues to leave alone? What makes the immune system go haywire and start attacking healthy nerve tissue, causing autoimmune diseases like multiple sclerosis? How does the immune system retain a "memory" to recognize and attack invaders years after an initial exposure?

Nobody really knew.

After working to answer these questions as a graduate student in the 1960s, Hood developed into a leading immunologist over the next two decades—before he became known as the man with the gene machine. "Lee was unquestionably the preeminent molecular immunologist in the world," said Perlmutter, now president of Merck Research Laboratories. Mark Davis, a former student of

Hood's who himself became a leading immunologist at Stanford, agreed. "Lee was the absolute intellectual leader of immunology from the '70s through the '80s," Davis said.

It was on this work that Hood competed for the top prize in science: the Nobel.

When Hood was a graduate student in the 1960s, his chosen field of immunology was full of mystery. Scientists were groping in the dark, unsure how best to gather data even to start tackling the hard problems. Yet there was action to be had in immunology. The decade before Hood dedicated himself to the field, Jonas Salk developed a vaccine against polio, a disease that crippled and killed young people. This was among the great triumphs of twentieth-century science. Vaccines worked, increasing life expectancy and quality of life. Scientists just didn't know *why* they worked, and why some experimental vaccines didn't provoke an immune response at all.

But it was clear that studying immunology promised significant payoffs for both health and medicine. Millions of people suffered from chronic and debilitating autoimmune diseases such as rheumatoid arthritis, multiple sclerosis, and lupus. A deeper understanding of immunology could pave the way for new and better vaccines, new and better drugs. It could improve surgery success rates, because doctors would know how to prevent the body from rejecting transplanted hearts, kidneys, or bone marrow. Especially optimistic researchers went further: Could you "teach" the immune system to turn its killing firepower toward cancer cells, seeking and destroying them just like an invading virus? The research questions were so ambitious that some scientists thought they might never be answerable. The tools to tackle them didn't exist. Neither did formal departments of immunology at most universities. The jargon of immunology was so thick, it intimidated many people in other biological disciplines.

But Hood could easily foresee an entire career's worth of grand adventures in immunology. One of his heroes at Caltech, Ray Owen, had done pioneering work in cattle twins that laid the foundation for the concept of "immune tolerance." Owen was a postdoctoral fellow at the University of Wisconsin in the mid-1940s—the "dark ages" of immunology, as geneticist James Crow later wrote. Owen raised the provocative possibility that the immune system was able to distinguish between foreign invaders to kill and healthy tissue to leave alone, or "tolerate." Owen's original manuscript was never published. A short version published in 1945 drew little notice at first, Crow wrote. But progress in immunology soon followed. Like any field of science, success begat more success. Ambitious young scientists like Hood were attracted. He had an opportunity to work with one of the founding fathers of the field. And Owen happened to be a kind soul willing to help students in need of career guidance. Hood was in.

But one couldn't just dive into the grandiose problems of immunology. Scientists of Hood's generation realized they first needed an intimate understanding of the structure and function of proteins—the molecules that carry out a lot of basic immune functions. Among the most interesting warrior molecules: antibodies. Antibodies are Y-shaped proteins, secreted by B cells, a type of white blood cell that comes from the bone marrow. Antibodies were fascinating because they appeared to act like heat-seeking missiles against invading pathogens. Each unique pathogen requires a distinct antibody to destroy it. The human body has a finite list of tens of thousands of genes, yet a seemingly unlimited capacity to produce antibodies—billions of them—exquisitely tailored to their killing task. Scientists had some working theories on how this was possible. But mostly they were stumped.

One working hypothesis was that the B cells became genetically altered after conception, in response to a person's environment. The key to antibody diversity, the thinking went, was through these post-conception DNA alterations, otherwise known as "somatic mutations." Another working idea was that B cells—some with genes inherited from the mother, and some with genes inherited from the father—swapped DNA with each other as needed and then reshuffled it to form the components of antibodies. These were "germline" mutations. "Just as you have 26 letters in the alphabet forming whole libraries of books, or 52 playing cards making up innumerable poker hands," the cell's ability to shuffle genes explains how the immune system can mount such an amazingly diversified antibody defense, as Dr. Thomas Waldmann, chief of the metabolism branch of the National Cancer Institute, later explained it in the *Washington Post*.

As a graduate student under Dreyer, Hood helped refine this hypothesis. He worked on showing that two genes could rearrange their DNA to provide instructions for making a novel protein, though he wasn't the first to definitively prove it. The idea ("two genes, one polypeptide") challenged decades of dogma that said only one gene is responsible for carrying the instructions for making a distinct protein ("one gene, one protein"). The answer eluded scientists for many years, in part because antibodies are large, complex molecules with subunits made of long peptide chains. These peptides came in a couple varieties, known roughly as "heavy chains" and "light chains," which come together to give the Y-shaped antibody its structure. Both chains had variable regions, found on the tips of the Y-shaped molecule, and these are the parts that bind with antigens. They were obviously important. It wasn't obvious, beyond that, where to look for clues to the source of antibody diversity.

This mysterious source of antibody diversity continued to fascinate Hood in the early part of his career on the Caltech faculty. Technology was improving in the 1970s, with the advent of manual DNA sequencing methods developed by Fred Sanger and Wally

Gilbert, but progress was still slow for Hood's taste. Lacking bullet-proof data, Hood equivocated in a 1978 paper published in *Nature*. It was titled "Rearrangement of Genetic Information May Produce Immunoglobulin Diversity." That wasn't very satisfying. Hard, convincing evidence was necessary to win over skeptical peers.

Hood's drive to develop the automated protein sequencer, and later instruments like the automated DNA sequencer, were influenced in part by this problem. He understood what could be done with powerful new tools that could throw off large volumes of experimental data. Rather than defensively dig in his heels, like many protein-focused biologists did in the 1970s, Hood was savvy enough to see where the action was heading. He knew he would have to break out of his comfort zone and embrace the new DNA-based molecular biology, or else risk falling into irrelevance. Linking DNA and proteins together—truly connecting the dots—would give rise to a more unified view of biology. Hood's deep knowledge of immunology and of biology's big unanswered questions are what enabled him to direct the way new automated protein and DNA sequencing machines should work.

The instruments that resulted were such a significant advance in the field of genomics that people eventually would forget Hood's reputation as an immunologist. He couldn't build the machines himself, but that didn't really matter. As Mike Hunkapiller, the leader on the technology side of the Hood lab, put it:

> He wasn't, and still isn't, a technologist, despite the reputation. That's not what he did. His chemistry skills were not that good, in a sense. His experimental skills, I don't think, were that good. He wasn't necessarily that close to the details, the practical ways, to get things done. But that's OK. [He] could bring people in who got the vision, and

he could [provide] the guidance to help nudge along
what needs to be done. ...

There wasn't a history at that point of building
automated technology to help solve these problems.
Lee's genius early on was in figuring out that we
not only need to deal with the qualitative aspects,
in getting enough sample to run an experiment,
but we're going to need to generate a lot of data to
solve some of these problems. It wasn't something
that was going to be scalable with just more pairs of
hands.

Hood wasn't the only one thinking about the new tools of biology,
and how they might be used to answer the questions about antibody
diversity. Through the mid- to late 1970s, it came down to a three-
horse race. There was Hood's team at Caltech, Phil Leder's team at
the NIH, and Susumu Tonegawa's team at the Basel Institute for
Immunology in Switzerland. Each was working independently on different pieces of the puzzle. "Lee Hood wanted to win the Nobel Prize
for sure. He worked like hell," said Bill Dreyer, the Caltech professor,
in a 1999 oral history interview. "He had [to do] more and more work
on antibodies to build up his image in the field. And he did—he did
superb work. This is a wonderful thing, that people are this way."

Tonegawa wanted the prize, too. Like Hood, Tonegawa had a
prodigious work ethic. Members of the Hood lab were fearful of
the Japanese scientist they saw as a "samurai warrior." Apparently,
Tonegawa told his first wife they couldn't have children because
it would be a distraction from science. He worked all night and
expected his colleagues to do the same.

Besides his drive and intellect, Tonegawa had an advantage the
others couldn't match: location. The United States had banned the

use of an important technique, and Tonegawa wasn't working in the United States. Early recombinant DNA techniques in the mid- to late 1970s created thrilling possibilities—along with ethical concerns. The techniques made it possible to splice DNA from one organism, like a human, insert it into a host cell, like a bacterium, and produce many copies of a functioning human protein. Fears that scientists might end up introducing a cancer-like virus in people led to a moratorium for a couple years, which stalled work in the United States.

No such moratorium existed in Switzerland. The recombinant DNA techniques were precisely what Tonegawa needed to get a jump on the long-standing question about the source of antibody diversity. "He was the smartest of all," Dreyer recalled in his oral history. "You couldn't use those tools yet in this country—they were banned—but he went to Switzerland."

The race was on. Given the importance of antibodies, the scientist who could pin down the source would probably win the big prizes.

The Albert Lasker Basic Medical Research Award was one of the biggest. It's sometimes called the "American Nobel." The prize, first granted in 1946, is announced each September. Enough time must pass that a scientist's contribution can be considered in the proper context. Hood was nominated in 1987.

Hood was excited. At age forty-eight, this would be the most prestigious award of his career. Just as enticing, the winner of the Lasker would be first in line for the Nobel Prize. A total of eighty-six Lasker Laureates have gone on to receive the Nobel.

When the judges' decision came down, Hood, Leder, and Tonegawa all shared the Lasker Award. Sharing the prize, then and now, was a common way for big prizes to honor the contributions of several scientists without going into the convoluted and often controversial business of determining whose work was truly first, and whose was most important.

The gene-swapping theory turned out to be correct. The theory of somatic mutations—in which antibodies took advantage of DNA variations in response to environmental stimuli after conception—was partly correct, too. "The real answer turned out to be a hybrid of both," Hood said. Antibody diversity had been solved.

Tonegawa's tour de force was unmistakable. He took advantage of some of new manual techniques for DNA sequencing and analysis. "In one crucial experiment, Dr. Tonegawa determined the genetic mechanism of antibody diversity and showed that new genes are created through rearrangement of DNA during B-cell differentiation. In further research, Dr. Tonegawa confirmed and determined the details of his discovery by cloning and sequencing antibody genes," the Lasker Foundation wrote. Years earlier, Bill Dreyer and his plucky graduate student Lee Hood had developed and refined a theory, which is critically important, but Tonegawa had confirmed it with hard data.

Leder was honored by the Lasker jury for "a stunning series of experiments" finding that when antibody-producing B cells perform their genetic rearrangement, it can sometimes lead to a disturbance in the regulation of cell division, giving rise to cancer.

The jury cited Hood for documenting "in elegant detail the immune system's method of rearranging pre-existing sequences of DNA to make the genes for each new antibody." As the *Washington Post* later put it, Hood and Leder "proved that the genes containing instructions for the protein chains that make up an antibody are spliced and shuffled as a B-cell develops, much as a railroad controller uncouples and recouples the cars of a freight train." The Hood team used molecular biology techniques to study the components—both the "light chain" and the "heavy chain"—that come together to make up antibodies. In describing what happened in the heavy chain, the team found an additional variable that contributes to antibody diversity, according to the Lasker Award jury. Tonegawa had been first and did crucial work on the more complicated genetic rearrangement that occurred in antibody light chains. But that wasn't the whole story. On the antibody heavy chains, Mark Davis said, "we really made a killing."

Hood was more than ready to take a victory lap. He flew to New York to accept the award. Eran, then attending Harvard, joined the celebration.

Back at Caltech, the Hood lab was buzzing with speculation. The boss was rumored to be a Nobel contender. If he won, they would all, by extension, bask in some of the glow that makes scientific careers. Which of the three labs had made the most important contribution? It didn't matter. The Nobel Prize can be shared by as many as three scientists in a given year. Since Hood, Tonegawa, and Leder shared the Lasker, they all could soon be jetting off to Stockholm.

Hood was ready for the fabled early-morning phone call in October 1987. But the phone didn't ring. While visiting Bell Labs on the East Coast, Hood got the bad news. Tonegawa, alone, had captured the Nobel Prize in Physiology or Medicine. It was only the second time in more than twenty-five years that the committee gave the award to a lone recipient. The Nobel committee citation said Tonegawa "completely dominated this area of research" from 1976 to 1978. "If you are given the Nobel Prize alone, what higher tribute can you be paid?" the immunologist Hans Wigzell of the Karolinska Institute in Sweden told reporters. Wigzell was a member of the Nobel committee that year, and given his expertise in immunology, he was surely an influential voice in the deliberations. When reporters asked Leder and Hood whether they had been unfairly snubbed, both were diplomatic. Both men, in separate interviews, told the *New York Times* that Tonegawa's prize was well deserved.

Years later, Hood called the loss "the most disappointing moment in my life to that point."

Each side had a legitimate argument. And, as with many major prizes—including the Academy Awards, football's Heisman Trophy and journalism's Pulitzer Prizes—there's a long history of behind-the-scenes campaigning and controversy surrounding the biggest award in science. Hood said he never found out why he lost. "The wife of a friend who was on one of the Nobel committees told me at the time that I should have more actively campaigned, and that perhaps would have changed the outcome," Hood said.

Hood wasn't that naïve about Nobel Prize politics. It was common knowledge that publishing groundbreaking papers and hosting stimulating seminars were just part of a winning prize campaign. It was also important to find one's way to Sweden, off the beaten scientific track, and give stirring lectures there. Once the antibody diversity work had wrapped up in the early 1980s, postdoc Mitchell Kronenberg recalled, Hood "made more than the expected number of trips to Sweden."

Why did Hood fall short? Years later, Wigzell offered some perspective. An immunologist about the same age as Hood, Wigzell recalled having "several very early discussions" with Hood about how the immune system recognizes invading antigens while at a Cold Spring Harbor symposium in the late 1960s. "Lee has made several very nice contributions to the question of especially B-cell cognition/recognition of antigen. I remember ... Lee being profoundly excited via Dreyer's thoughts." So that much was clear to Wigzell: Dreyer had the crucial insight that genes might reshuffle their information to form antibodies. Hood, as a graduate student, discussed and refined this theory with his adviser and later, as an independent investigator, ran experiments that shed light on parts of the antibody diversity mystery. His instruments made it possible to drill deeper into fundamental questions of molecular immunology. But Nobel Prizes, at least in the physiology-or-medicine category, typically are not given for inventing technologies. "One important thing to note ... that is sometimes not given due weight, is that the [Nobel Prize in Medicine] should be given for a DISCOVERY whereas for instance for Chemistry it could be an IMPROVEMENT," Wigzell wrote to me in 2014.

So if the Nobel Prize for Medicine is discovery-driven, not technology-driven, then the only realistic way Hood could win was for the quality of the results from his experiments in immunology. Looking back, Wigzell said:

> It is of course now quite clear that molecular biology tools were drivers in the development of a deeper understanding in immunology. Here, Lee was instrumental in … understanding the way biology would move, thus putting great force into the development of pioneering instruments for protein and DNA sequencing and synthesis. This may be his most important contribution to science.

> Being an immunologist by training, Lee was also then quite successful in the analysis of the various key molecules in the mammalian immune system, [including] the immunoglobulin molecules and their functions in the immune recognition and specificity. Relatively speaking, his contribution here is truly excellent. But maybe not as equal in consequence to the development of medical science as the technological achievements.

Many other accomplished scientists and inventors never had their moment of glory in Stockholm. Thomas Edison, inventor of the lightbulb; Tim Berners-Lee, pioneer of the World Wide Web; and Stephen Hawking, the physicist who explained the death of black holes, are among the illustrious people who never got the Nobel.

Tim Hunkapiller recalls having dinner with Hood after he flew back to Pasadena. "It was the only time he was shocked at how disappointed he was," Hunkapiller said. "He didn't anticipate it affecting him." Not long after, he was philosophical about it. During a long drive to a scientific retreat, Roger Perlmutter said, "I remember him saying, 'I didn't get into this business to win prizes.'" Still, Hood couldn't help feeling snubbed. "He felt he deserved it, it had been an important problem, and he made seminal contributions," Perlmutter said. The Hood team respected Tonegawa's rightful claim to a Nobel. But many thought that he shouldn't have gotten the prize all by himself.

Antibody diversity was solved. The next great race was on. Two even more mysterious areas of immunology beckoned: T cells and the MHC. T cells, in their various subtypes, performed essential roles in the immune system's orchestra. The MHC—the major histocompatibility complex—was a series of molecules that bind to the surface of an antigen and display bits of its proteins so that certain T cells can recognize the pathogen for attack. They act sort of like matadors waving bits of red flag in front of charging bulls. Scientists discovered a phenomenon called MHC restriction; it governed when the immune system could recognize a foreign antigen. If you could control that, you could control whether the immune system recognized and rejected an organ transplant. With that, scientists had a richer picture of the immune system. Antibody diversity was only part of the story. If you really wanted to know how the body develops an immune repertoire, then you also had to know how T cells form receptors on their surface that bind with invading antigens.

By the mid-1980s, the Hood team had plowed through the complex genetic underpinnings of antibody diversity, the T cell receptor, and the MHC.

Research on antibody diversity and the MHC was "absolutely pathbreaking," said Ellen Rothenberg, who became a longtime

Caltech faculty member. Irv Weissman called the work, along with the development of the automated DNA sequencer, one of the central achievements of Hood's career.

Taken together, this work created a deeper understanding of what is now known as the "adaptive immune system." This system provides the body's repertoire for learning to fight antigens and remembering how to fight invaders it has seen before. The idea is essentially this: the immune system, through complex genetic rearrangements, evolves to keep up with the relentless and rapid onslaught of invaders that display all kinds of complex structures: microbes, chemicals, and even insurgent cancer cells. Seeing the immune system adapt quickly to stimuli forced many scientists to think deeply about evolution.

Hood remembers this time as a "golden era," when he was at the peak of his career. He had recruited a cast of young stars in immunology and technology, and they were advancing quickly on both fronts.

Science, of course, isn't about one triumphant discovery after another. There are a lot of dead ends.

Michael Steinmetz was familiar with that harsh reality. He was Hood's point man overseeing fast-moving work on the genetics of the MHC. Steinmetz's team focused on sequencing a genetic region known as the I-J locus, which was widely believed to hold the code that gave T cells their ability to recognize specific pathogens. Labs all around the world were forming hypotheses, gathering data, and refining their hypotheses as part of this quest. Intense scrutiny was being applied to this specific zip code of sorts on the genome. Mark Davis recalls what happened next:

> There was this very well-defined locus, called I-J.
> It was supposed to be there in the middle of the
> MHC, and was the key to a lot of work on T cells.

[Steinmetz's crew] raced through and found nothing.
… It was a huge psychological shock to the whole
field of immunology. They put a lot of effort into it,
because they thought this was going to be it. Then it
went away.

Knowing where the code *wasn't* was useful, certainly. But where was it? Finding out was important, because then researchers could start to look at commonalities and abnormalities in the code that might provide valuable clues for medicine. "The identification of the complementary DNA encoding these chains was the real Holy Grail yet to be found," wrote Arthur Weiss, a UC San Francisco researcher, in a 2005 review article. Labs all over the world were pouring millions of dollars into this quest, throwing lots of eager young people at the problem.

Mark Davis was one. As a grad student at Caltech in the '70s, Davis chafed under his first adviser, Eric Davidson. Looking for a fresh start in 1977, he sought out Lee Hood. The two were a good match. Davis thrived on autonomy. Hood needed self-starters. Hood's expertise was still confined mostly to protein molecules. Davis had a growing set of skills in nucleic acids—DNA and RNA. Davis's skills improved, aided in part by the technical chops Caltech imported when Tom Maniatis, a pioneer in gene isolation and cloning, joined the faculty in 1977. Under Hood, Davis finished his PhD work by the end of 1980 and took a postdoctoral fellowship at the NIH. He was poised to apply his genetic engineering skills to the T cell receptor problem.

The theory was that, like antibodies, T cells rearrange genetic material to form receptors that could bind with invading antigens. T cells have thousands of these receptors on their surface. But none was available in large quantities. That made it hard to get enough data to prove the theory with statistical confidence. Davis knew, from his graduate school days working for Hood, that getting the DNA code was going to be essential to getting that Holy Grail.

This was such a hot area of science that Stanford offered a faculty job to Davis in 1983. Until then, he was just a young investigator with a small team at the NIH. But he had devised a novel approach using mice to find the genetic basis for the T cell receptor.

Word began circulating that Davis was onto something big. He began to hear competitive footsteps. Davis learned through the grapevine that Hood was mobilizing resources to beat him in the T cell receptor race. It was not unprecedented in the rough-and-tumble scientific world, for a thesis adviser to try to chase down a former student. But it was unusual. More common would be to give encouragement or attempt to form a partnership. Hood did neither.

Davis thought he had a comfortable lead. He knew that Hood's lab didn't have his level of expertise in DNA or his novel method. But he also knew he had fewer resources than Hood, with the limited budget of a newbie, and he couldn't afford to be complacent. He needed to establish himself as the leader in the field. So in August 1983 at a scientific conference in Kyoto, Japan, he rather brazenly declared—without showing hard data—that he had done it: isolated a messenger RNA that encoded a component of the mouse T cell receptor. What Davis didn't know was that an obscure researcher at the University of Toronto named Tak Mak was working on the same problem, having come across what he believed to be vital DNA code for making the T cell receptor in human cells. Mak was shocked to hear word of Davis's presentation in Kyoto. "We were jolted out of our complacency and naïveté," Mak recalled. He put a second graduate student on his project. Speculation mounted about who had done what. A month or so after Davis gave his talk, Mak called Davis and said he had identified the T cell receptor's beta chain in human cells. He asked how they might work together. Davis remembered the conversation this way:

I asked, "Does it rearrange?" He said, "I don't know."
I thought, "Who is this guy? Is this some kind of
crank caller?"

I thought I was being generous. I said, "If you send
me your sequence, I can tell you if it's the same as
ours. He said, "I was thinking *you* could send me
your sequence." I said, "No dice."

Davis hung up, wondering whether Mak was bluffing. By
December, Davis couldn't ignore the gossip: Mak was onto something, and he had enlisted the formidable Hood lab as a partner to
extend on his initial findings. Then, in January 1984, Irv Weissman
asked Davis whether he had submitted his T cell receptor paper to
a scientific journal yet. "I'm still fussing with it," Davis replied. He
had wanted to make a grand splash with two papers at once. The
exacting brand of scientific perfectionism required for such a tour
de force had its merits, but it could also be slow. Lee Hood could eat
slow scientists for breakfast. Weissman gave his Stanford colleague a
warning: "Now is not the time to get writer's cramp," he said.

Davis felt stung. He hadn't been personally close to his former
adviser, but he had done good work for the boss. He didn't see this
coming. Instead of staying out of it and letting Davis have his day in
the sun, Hood had partnered with a rival. Was Hood really so fixated on getting wins that he was willing to crush a former student's
career along the way? Hood already had a reputation for picking up
ideas from others, assimilating them into his efficient lab, improving
them, and gaining the competitive advantage. He sometimes told
people in his lab that he didn't care where the good ideas came
from, he just wanted them all. Some joked (or complained) that the
Hood lab operated like the Borg alien race from *Star Trek: The Next
Generation*. Their famous line: "We are the Borg. Lower your shields

and surrender your ships. We will add your biological and techno-logical distinctiveness to our own. Your culture will adapt to service us. Resistance is futile."

Loyalists to the Hood lab scoffed at such notions. Davis was hypersensitive about ordinary scientific competition, they said. But even decades later, Davis remembered how rattled he was at the time. Fear of being scooped by the all-powerful Hood, real or imag-ined, drove him. This was a career-making discovery for a young scientist. He couldn't afford to finish second. "What's striking to me is that people would say things like, 'I hear Lee is getting into this T cell receptor work; what are you going to do now, Mark?'" Davis said in a 2013 interview. "They thought I didn't have a chance."

He rushed to finish his two papers by February 1984. He drove them to a DHL office at the airport. He arranged to have a courier pick up the envelope at London's Heathrow Airport and whisk it to the offices of *Nature*, the top scientific journal. Eight days later, Davis got a call from an editor. His papers had been accepted. It was an astonishingly quick turnaround; publication typically takes a couple of months as editors wait for peer-reviewed comments and revisions.

Davis showed that T cell receptors have genes for variable, con-stant, and joining regions—quite similar to the genes that provide the code for making antibodies. It was a critical finding that estab-lished how T cell receptors recognize and combat invaders. The same March 8, 1984, issue of *Nature* also published Mak's work on the T cell receptor. The editors had been holding Mak's paper—which contributed key knowledge about human cells—since November. The trio of publications in the same issue from two independent labs made a convincing case.

The discoveries skyrocketed Davis and Mak to international acclaim. Hood's name didn't appear on that seminal batch of T cell papers, since Mak had completed the work before his partnership with Hood. But between June and September 1984, Hood and Mak coauthored five papers on T cell receptor genetics in mice and humans, published in *Cell* and *Nature*. Davis couldn't rest. He published five more papers in *Nature* that year as well. "It was over the top. Lee is famously competitive," Davis later told the *Wall Street Journal*. For at least the next three years, Hood's team on the T cell receptor, led by Mitchell Kronenberg, continued experiments to answer remaining questions. Kronenberg doesn't recall Hood ever openly expressing a desire to "crush" Davis, despite rumors to the contrary.

Hood's collaboration with Mak was brief. The agreement was that Mak would work on human cells and Hood would stick to mice. But Hood wasn't the type to stay in a small sandbox for long. His aggressiveness made many collaborators leery. "He already had a reputation of being someone who would take everything you've got and then say 'See you later.' He wasn't looking for a long-term relationship. You might be a coauthor on the paper, if you were lucky," Davis said.

What's the old line about nice guys finishing last? After all, many top scientists got to the top because of their sharp elbows. James Watson, the co-discoverer of the DNA double helix, was famously fearful of Linus Pauling and dismissive in his treatment of Rosalind Franklin, a collaborator. As Davis said, "I don't think it's the first time it's ever happened in the universe, but it's unusual … especially when it was your own student."

The work on T cell receptors had been laborious. DNA sequencing in the mid-1980s was painstaking and prone to human error. One gene might be composed of a few hundred, or a few thousand, base chemical units of A, C, G, T in a unique signature. It might take a graduate student's entire career to nail down the DNA sequence for just one gene. Sequencing the adrenaline receptor took a full decade. It reminded Hood of his graduate-school days, working on protein extraction and purification. But this was the leading edge. Hood started to turn his attention away from immunology. Some of the big questions had been answered; the others remained far beyond the reach of existing technologies. It was time to focus on a greener pasture: genomics.

For most of a decade, Hood had been dreaming about what he could do if he could sequence DNA with a truly high-speed, large-scale, automated set of instruments. Hood was eager for instruments to handle the drudge work of sequencing and for computers to crunch the data. Biologists would be able to ask bolder, broader, data-rich questions—to imagine hypotheses they never would have dreamed up before. The vision struck many as pure science fiction. The boldest idea was to obtain the first complete human genome, with its staggering row of six billion base chemical units of DNA. Some thought it would take a century.

Even Hood had his doubts about the Human Genome Project the first time he heard about it, in the mid-1980s. It didn't take him long to change his tune. First, he just needed a better machine.

Chapter 10

The Invention That
Sparked a Revolution

"When we did the DNA sequencer, I kind of knew it was important," Hood said years later. "I didn't know it was going to be historic."

The invention was comparable to that of the printing press, which enabled mass production of books that spread ideas fast and far. In giving rise to the genomic information age and a whole new way of understanding human biology, the DNA sequencer was "arguably one of the most important advances in biology in the 20th century," a US District judge wrote in a decision that settled an intellectual property dispute.

Starting with Hood's raw, barely automated prototype in 1986, new generations of DNA sequencers continually improved at a rate without precedent in human history. The machines made gains even faster than computers did under Moore's law, the visionary 1965 forecast that computing power would double every two years.

By 2015, instead of spending years laboring to determine the order of DNA units making up a single gene, it was possible for a grad student to push a few buttons and have a gene sequence in seconds. The machines didn't stop there. They could roar through all six billion chemical units of DNA in a complete human genome. The first genome had taken thirteen years to complete and cost about $3 billion. A little more than a decade later, the work could be done in *one day* for roughly $1,000. The result? A tidal wave of genomic data.

The machines pumped out so much genomic data that researchers didn't know what to do with it all—or how to interpret it. A whole new scientific discipline, computational biology, was born. It would take years to sort through the genetic code and decipher what it meant in terms of health and disease.

Once DNA sequencing tools became widely available, though, doctors and researchers quickly gained new ideas for solving mysterious illnesses. A six-year-old boy from Wisconsin named Nicholas Volker owed his life to DNA sequencing. Nicholas had dangerously inflamed intestines, for which he endured more than a hundred surgeries. Then, in 2010, his doctors tried something unusual— they sequenced his DNA. They found a rare mutation. The information suggested a new path forward: a bone marrow transplant from umbilical cord blood. It saved Nicholas's life. The *Milwaukee Journal Sentinel* wrote a Pulitzer Prize–winning series about the boy, and the hopeful narrative spread from the scientific literature to *The Today Show*. Scientists, doctors, and parents of other desperately ill children turned to DNA sequencing.

But long before all of that, in the early 1980s, automated DNA sequencing was just an idea.

The vision for the future of biology went like this: build machines that could both read and write in the fundamental language of life. There would be four kinds of tools: DNA synthesizers,

DNA sequencers, a protein synthesizer, and the original protein sequencers. The family of instruments working in harmony would churn out biological information around the clock. Although Hood often described this central "microchemical facility" as the result of "his" vision, he wasn't a lone genius. His former adviser, Bill Dreyer, was a coauthor of the paper that sketched out the plan. So was Mike Hunkapiller, along with collaborators at the University of Colorado, Abbott Laboratories, and the Jet Propulsion Laboratory. Hood detailed this idea for four integrated instruments and submitted it in 1982 to the journal *Nature*. Skeptical, the editors held publication for two years. First, they wanted to see realistic prototypes of an automated DNA synthesizer and an automated DNA sequencer— the final pieces of the puzzle.

The synthesizer came first. Making DNA through synthetic means became an everyday essential task in many labs. DNA synthesis, for instance, became crucial for amplifying tiny DNA samples so that they are easier to study. Criminal forensics labs in later years would use the polymerase chain reaction techniques to identify criminals, or exonerate suspects, based on trace DNA samples in hair, blood, or bodily fluids that needed to be amplified.

Because Hood wasn't an expert in DNA himself, he needed to lean on up-and-coming postdocs and peers. One such expert was Marvin Caruthers, a chemist at the University of Colorado. He had developed new phosphorus-chemistry methods for synthesizing DNA by hand in 1980 and 1981. Hood imagined the new technique could fit in his grand plan. With Caruthers's permission, Hood sent one of his deputies, Suzanna Horvath, to the Caruthers lab to learn the DNA synthesis technique. She brought it back to the Hood lab so the team could attempt to automate the process. In Hood's telling, Caruthers was willing to teach Horvath but didn't really see the point of automation. "Marvin argued that he could easily teach the procedure in a week to anyone who needed to synthesize DNA," Hood wrote years later, "and because he felt the demand for oligonucleotides [short DNA and RNA molecules] was never going to be significant, why bother with automation? I convinced him otherwise."

Remarks like that irritated many of Hood's peers. With gusto, Hood would tell and retell these stories about how he was ahead of the curve while portraying other biologists as if they were members of the Flat Earth Society. Fellow scientists grumbled in private. Occasionally they set the record straight in public, portraying Hood as something less than biology's only wise man.

If Caruthers had doubts about the automation of DNA synthesis, it was only for the briefest of moments. He and Hood were both among the original scientific advisers to Amgen in 1980. They were in regular contact through the early 1980s, well aware of each other's work. Caruthers taught Hood's postdoctoral fellow the manual DNA synthesis methods in 1981. That same year, Caruthers joined Hood as a cofounder of Applied Biosystems, dedicated in part to automating DNA synthesis. Automation was a shared goal.

When Horvath returned to Caltech in 1981, Hood directed his team of mechanical and software engineers to build a prototype machine that could suck the base units of DNA out of containers and chain them together in the desired order of As, Cs, Gs, and Ts. But the project stalled. Horvath and Hood never got a patent for their work on an automated DNA synthesizer and didn't publish anything on their work until years later, in 1987. Applied Biosystems quickly grabbed the lead. On a roll selling its protein sequencer, the company was able to pour considerable resources into its next instrument: a robust DNA synthesizer, one that met exacting standards and could be manufactured consistently for sale. Their first automated DNA synthesizer was at least one hundred times faster than the manual method.

How much did Hood contribute to it? Accounts differ. There were striking similarities in the underlying guts of the Hood lab's protein sequencer and Applied Biosystems' DNA synthesizer. But the chemistry was quite different, and not Hood's forte. Hood insisted in later years that Applied Biosystems benefitted from a free-flowing exchange of ideas with the Hood lab. (Hood cofounded the company but never worked there.) When Caruthers offered his own narrative of how that "gene machine" came to be, he didn't cast Hood as the leading man. Caruthers's article, published three decades later in the *Journal of Biological Chemistry*, was titled "The Chemical Synthesis of DNA/RNA: Our Gift to Science."

Caruthers wrote that Sam Eletr, the CEO of Applied Biosystems, played the vital leadership role in jump-starting the automated DNA synthesizer work in 1982:

> Sam hired two or three scientists each from Lee's and my laboratories to design and build our first instruments. From my laboratory, Bill Efcavitch and Curt Becker were among the first employees (later, Lincoln McBride as well). Bill and Curt started with a series of valves, tubing, small HPLC-grade (high-performance liquid chromatography) silica columns, a tank of liquid nitrogen, and a large piece of plywood.

> Within a few months, they were synthesizing DNA on this platform. Sam asked John Bridger, an engineer hired from Hewlett-Packard, to design a DNA synthesizer, which became known as our 380A machine.

> By December 1982, Bill installed the first commercial synthesizer in my laboratory, and Applied Biosystems began shipping the instruments in 1983.

"Hood's group tried—largely unsuccessfully—to automate an early synthesis chemistry from Marv, but by the time they published anything, ABI had been selling instruments based on Marv's phosphoramidite chemistry for several years," said Mike Hunkapiller, the longtime Hood lieutenant who helped found Applied Biosystems, worked there twenty-one years, and witnessed the DNA synthesizer development firsthand. "Hood's group contributed nothing to these instruments or any of Marv's chemistries."

Years later, when questioned directly about his lab's role in the DNA synthesizer and confronted with evidence presented by others, Hood offered a more nuanced view. "Certainly, Applied Biosystems optimized the instruments," Hood said. "The only instrument they didn't have to do a lot of optimization on was the first one, the protein sequencer. For all of the rest, they played a major role in developing robust forms of the instruments." Even that interpretation by Hood baffled many of the early employees and investors of Applied Biosystems. They said the protein sequencer, like all the other instruments, needed considerable engineering before reaching true commercial-grade quality.

It is fair to say that Hood was a catalyst for the DNA synthesizer, but in the end, he wasn't its creator. Caruthers got most of the credit for blazing the trail in DNA synthesis. He won the National Medal of Science for the underlying chemistry that made it possible. Hood and Horvath's paper on their automated DNA synthesis work ended up in a narrow technical journal, *Methods in Enzymology*, in 1987, drawing fewer than three dozen citations from other scientists—a sign of modest impact. By that year, Applied Biosystems was sprinting ahead. The company generated $84 million in sales, and had grown to 788 employees.

It's not surprising that Hood wasn't ideally positioned to get in on the action in DNA. He was an expert in protein molecules, the ones that fold up into all kinds of odd shapes to drive most biological functions in cells. The frontier of biology was moving to DNA and RNA, the substances that hold the recipes for making all those proteins. It was extremely difficult to do what Hood was attempting to do, which was to be at the forefront of both protein chemistry and molecular biology.

Biologists like Lee Hood were on a quest: How do we decode fundamental information from nucleic acids? How is that information translated into the wondrous and astounding complexity of a human being—with trillions of cells working in concert, responding in real time to all kinds of environmental stimuli? The task was bewildering, the payoff grand. "The 1980s appear to be the decade of biotechnology, and two of its aspects—recombinant DNA and microchemical instrumentation—will impinge greatly on medicine of the future," Hood and Mike Hunkapiller wrote in March 1983. It would lay the groundwork for a revolution in medicine.

But not every biologist wanted to be on this quest. For one thing, many peers in biology were slow to embrace such a far-reaching vision. One day's exciting discovery is usually followed by the experiments that fail or run into dead ends for unknown reasons. The path to greater understanding in biology doesn't follow a straight line.

Partly because of all this, molecular biologists are skeptical and detail-oriented by nature. While meticulous thinking is essential to scientific progress and to weeding out crackpot claims, it can be taken to extremes. Scientists often develop a worm's-eye view focused on ever more esoteric questions. Some who heard Hood articulate his vision—a view from the sky above—dismissed it as sloppy or hokum. Many saw a threat to their comfortable ways of doing science.

Synthesizers and sequencers created a highly quantitative, data-intensive form of biology, which promised to make mathematics a more important skill. Many biologists, however, had pursued careers in biology partly because they *didn't* need to know as much math as physicists. The fear was palpable. People got hostile. Once, when Hood was giving a seminar in Woods Hole, Massachusetts—a bastion of traditional biology—his host declined to show him to his hotel.

This automated version of biology also promised to speed everything up. Scientists feared that a key part of their culture, the contemplative beauty and wonder of biology, would be lost in some kind of techno-jock mad dash.

Three decades later, many biologists would shake their heads at the resistance to change. Other fields of science, from chemistry to physics, regularly embraced innovative technologies. New tools could help gather new types of data, they understood, which could drive new discoveries. That was conventional wisdom throughout most areas of science. But it didn't gain broad acceptance in biology until the mid- to late 1990s. "This prejudice against technology was ridiculous," said Maynard Olson, a genomics pioneer who became friendly with Hood in the '80s before clashing with him years later. "Lee's view was rational, that this was the direction biology needed to go."

If you had instruments that could sequence and synthesize proteins and DNA, then you could look at not just one gene at a time, but many genes, Hood and others theorized. Further in the future— seemingly a galaxy away to scientists in the 1980s—it would be reasonable to determine the order of the six billion DNA sequences that make up a human cell. Even further in the future, it might be possible to compare complete genomes from tens, hundreds, or even thousands of people to look for subtle variations that lead to health and disease.

It would have been hard to imagine such a thing in a late 1970s biology lab. DNA sequencing was just getting started. Frederick Sanger in the United Kingdom and Allan Maxam and Walter Gilbert at Harvard University had just published their respective methods for sequencing DNA by hand.

Their methods were technically complex, required hazardous chemicals, and used radioactive labels attached to the DNA. Scientists used a small jolt of electricity to push the DNA molecules through a gel. Based on how far the DNA traveled through the gel, scientists could guess the order of its nucleotide units, A, C, G, and T. And it was a guess. They'd peer at X-ray film images to interpret the order of the DNA sequence. It was a long, tedious, technical process that was prone to error. Many bright, young grad students sought to avoid the work, or pass it off to more gullible youngsters.

About two-thirds of the Hood lab in the 1980s was dedicated to immunology research. For one thing, it was the more socially acceptable line of work in the view of his biology peers. Technology people still were considered outsiders, and second-rate at that. Hood's technology group struggled for years to win any financial support from the usual federal granting agencies. Immunologists needed to be fluent in the language of DNA, RNA, and proteins to answer basic questions about how the immune system works. The problems were scientifically hard and had important medical ramifications. Hood devoted most of his time and attention to immunology.

Yet even when the technology work sputtered at times, Hood stubbornly insisted that it must continue. Technology promised to help answer some of the hard questions of immunology. If the big-name scientists who populated federal grant review panels smiled on his immunology work and frowned on his technology work, then he'd find another source of support for his technology. He was relentless in his quest to raise private money. Johnson & Johnson. Monsanto. Upjohn. Procter & Gamble. All were among the twenty-seven foundations, government agencies, and corporations that poured financial support into the Hood lab during the 1986/87 academic year alone. It was more than double the number of sponsors for any other biology lab at Caltech.

All the instruments Hood envisioned were making strides in the 1980s. The DNA synthesizer was in production. Peptides would soon be synthesized from individual amino acids, which could be chained together to make proteins for experiments. By 1983, a talented Hood deputy named Stephen Kent spearheaded the peptide synthesizer project in collaboration with Applied Biosystems. Those technologies soon shot from novelty to workhorse industrial tools. The protein synthesizer, protein sequencer, and DNA synthesizer were no longer just a vision. They were increasingly becoming real, and impactful.

But DNA sequencing remained the ultimate prize. If you wanted to truly connect the dots between variations in the underlying DNA and in the proteins, you needed all four instruments operating in something reasonably close to harmony. The DNA sequencer, though, was technically the hardest thing to automate. DNA sequencing required chemistry expertise that the Hood lab didn't have. Not to mention expertise in lasers and optical readers. As Hood said in an oral history interview with the Chemical Heritage Foundation:

> That [DNA sequencer] was the instrument that was
> *truly* integrated with regard to technologies. We
> needed to do really good chemistry. We needed to do
> really good engineering, and we needed to do really
> good software to kind of get that thing to go. But
> these other things were duck soup by comparison.
> That's why it took us five or six years to get that done.

Henry Huang, a postdoctoral fellow in biochemistry with a reputation for creative thinking, came to the Hood lab in 1977. Two years later, Hood gave him the hard assignment to build a DNA sequencer. Huang devoted 15–20 percent of his time to the project, and he oversaw a couple of engineers who were building prototypes. Acting in his usual capacity as a sort of executive producer, Hood had raised private money to get the project started. Huang gave regular status reports to the boss. But by late summer of 1982, the money had run out. Huang decided to leave. He'd had a productive stint, publishing more than ten scientific papers in the Hood lab, and after five years as a postdoc, it was time to move on. He got a promotion, an assistant professorship at Washington University in St. Louis. The transition made for a good time to reassess how the DNA sequencing project was going. Hood wanted a different set of eyes to assess where things stood. He asked Lloyd Smith.

Lloyd Smith was a hotshot. He had earned his PhD in biophysics at Stanford University, with training in physical chemistry—an especially meticulous branch of science. Smith had worked under Harden McConnell, a distinguished chemistry professor, and gained some experience with instrument development. He had a productive graduate school career, with nine peer-reviewed papers. But Smith didn't know anything about molecular biology and didn't read much scientific literature outside his discipline. When it was time for him to find a postdoctoral fellowship, a friend suggested he contact Hood. By then, Hood was already famous and running one of the world's largest academic labs, with seventy people. Smith was only dimly aware of Hood's reputation. He sought him out anyway. Hood agreed to a meeting.

Like so many others before him, Smith was impressed by how intense and passionate Hood was about science, paired with his laid-back West Coast way. Smith had considered working at Rockefeller University in New York, a place where people wore ties to work and made science feel joyless. Hood was different.

Smith and Hood, plus Mike Hunkapiller, sat down for their interview at an Italian restaurant near Caltech. They ate pasta and talked about football—or tried. Hood asked what Smith thought about John Elway, then a star quarterback at Stanford. Smith, not a football fan, was tongue-tied. It hardly mattered. What Smith did know was synthetic chemistry, laser optics, electronics, and some basic computer programming—skills that would all be essential to designing an automated DNA sequencer. Even though Smith hadn't lined up any of his own funding for his postdoctoral pursuits, something Hood usually required, Hood found some money in his own budget to bring Smith on board in 1982.

Showing up at Caltech, Smith was treated with the usual indifference. He drove a beat-up 1964 Chevy Nova to Pasadena with all of his possessions. When he arrived, nobody knew who he was or why he was there. As usual, Hood hadn't bothered to inform the lab about the new recruit. No space or supplies were set aside to help him get started. Like everyone else, Smith would have to sink or swim. As was often the case, a lab veteran would take pity on the newbie and offer a helping hand. In Smith's case, that person was Joan Kobori. If Smith wanted guidance from Hood, he'd have to get in line. Hood had a voracious appetite for work, and he worked with blinding efficiency, but he still couldn't do everything at once. He was traveling constantly, writing grant proposals, firing off manuscripts to journals, raising money, managing (or trying to manage) a seventy-person lab, and handling faculty and policy issues as chairman of biology. Comforting nervous graduate students wasn't a priority.

Smith was the swimming type. He threw himself into learning molecular biology. As a chemist, he was an outsider in the Hood lab, but that enabled him to look at things with a fresh perspective. He didn't like what he saw. "I was told that a great way for a novice to learn molecular biology was to sequence a gene, so in short order I found myself in the trenches with everyone else, working sixteen-hour days doing the repetitive business of DNA sequence analysis," Smith wrote in an unpublished manuscript. "It did not take long for this to demoralize me."

So when Hood asked Smith to assess Huang's project, Smith was glad for the chance to make use of his expertise.

Smith trekked over to the other side of campus for a look. What he saw, he said, was a "Rube Goldberg" contraption. The prototype used a large parabolic mirror to focus ultraviolet light onto DNA samples and then read the DNA based on the rate of light absorption by each of the four base chemical units. After doing a simple calculation he knew from chemistry graduate school, Smith concluded that such an approach would never be sensitive enough to work. He recommended that Hood pull the plug.

Hood, as usual, saw failure as just an opportunity to talk about more promising technical approaches. Lots of ideas began circulating through the lab in the fall of 1982 about how to do automated DNA sequencing. Smith suggested attaching fluorescent tags to the bases of DNA, because they would be three orders of magnitude more sensitive to optical detectors than ultraviolet light.

Meanwhile, Tim Hunkapiller, whom Smith considered a creative thought partner, shared an intriguing article from the May/June 1982 issue of *High Technology* magazine. It described a novel gas chromatography system that had been sent into space. Chemists were using the instrument to separate and analyze compounds. The article depicted a spiral-shaped capillary that could be meters long, but scribed onto a tiny silicon wafer, a few centimeters in size. Could a capillary tube be used to separate and analyze DNA? Tim Hunkapiller discussed it with his brother, Mike, and Lloyd Smith.

Quickly, the Hunkapiller brothers concluded it would be too difficult or impractical to put the usual four separate gel lanes, one for each of the four DNA base units, onto a silicon wafer. "We were stumped at first," Mike Hunkapiller said. Smith raised his concerns about the ultraviolet-light system. Again, he suggested the spectrally distinct fluorescent dyes. The idea of fluorescent dyes was important not only because of the sensitivity, but also because it provided an alternative to the dangers of radioactivity. And it would allow one lane instead of four.

In a meeting on October 1, 1982, Mike and Tim Hunkapiller, Smith, and Hood agreed to move ahead with a four-color fluorescent tagging system, running the DNA through a single capillary lane. Each color would represent a different base unit of DNA.

Mike Hunkapiller advocated the approach. It would be technically harder to do—getting four different fluorescent dyes to attach to the DNA bases without altering the DNA, all while sending off distinct colors that an optical detector could read. But if you could run the DNA samples through a single lane with four distinct colored tags, and if you could program a computer to read the data output, then you had something. "Mike was excited. Tim was excited. There were a lot of good vibes on this four-dye idea," Smith said.

According to an invention disclosure form filed with the Caltech administration, Lloyd Smith, Mike Hunkapiller, and Tim Hunkapiller conceived of a new automated DNA sequencing machine on October 1, 1982. Lee Hood was their prime witness. But there's no immediate record of that momentous meeting. The invention disclosure was based on Smith's notes dated October 29, 1982. The form didn't get filed with Caltech until more than a year later— November 28, 1983. "The necessary chemistry has been developed: the design and construction of a first prototype is in process," the inventors wrote in the disclosure form, a crucial first step toward filing a patent. "[I]t will enable much more rapid, reliable, and inexpensive determination of DNA sequence."

Either it wasn't immediately obvious that this particular process constituted an actual invention, or the principals were slow in getting their paperwork together to seek a patent.

With renewed enthusiasm, Smith went back to the bench to work on the DNA sequencer part-time while doing other projects. The first job was to develop the chemistry to enable him to attach fluorescent-dye tags to the DNA without damaging or altering the DNA. Within weeks, he started to get discouraged. There was the technically daunting nature of the problem. He also perceived a lack of support for this work within a lab dominated by immunologists. Smith went to Hood. "Maybe I shouldn't work on this," he said. Hood's optimism was just what a young postdoc needed. Years later, when he was a professor at the University of Wisconsin, Smith said:

> He gave me a big pep talk. I even think about
> it today, when I give my students pep talks. He
> said, "Oh no, this is VERY important. This would
> be PHENOMENAL if you can do automated
> sequencing." He was incredulous that I didn't realize
> how important it was, and that I'd even say that.
> That was good. That was him doing what a mentor
> should do, steering people in the right direction. I
> got pumped up and went to work on it some more.
> That was a very important, positive contribution
> from Lee.

Once Smith had figured out how to attach tags to the DNA units, he needed to find dyes that threw off distinct light signatures that could be consistently and accurately read by optical technologies. And he needed some help. Hood, always the golden-tongued fund-raiser, secured more support from the Donald E. Baxter Foundation. That enabled Smith to hire a small team, composed of Jane Sanders, Rob Kaiser, Peter Hughes, and Chris Dodd. Eventually, more private money for the project flowed from agriculture giant Monsanto, the pharmaceutical company Upjohn, and the Weingart Foundation.

In a letter to the president of the Baxter Foundation in 1984, Hood provided an update:

As I told you over the phone, we feel that the postdoctoral fellow working on this problem, Lloyd Smith, and his technician, Jane Sanders, have made absolutely remarkable progress on this problem over the past year. Where we have to go is absolutely clear. In another two years, we hope to have an instrument that will revolutionize molecular biology. I want to thank the Baxter Foundation for its support over the past year and I hope that our needs for this next year are not excessive.

At the time of the report, Smith was wrestling with tough technical hurdles. Getting the right fluorescent dyes to throw off a distinct signature for each base unit of DNA was time-consuming, difficult work. Three years into the project, Smith had no published paper to show for the work, nor a working prototype. He was exhausted. He wanted to write a paper and be done. "Lee, is this data good enough?" he asked.

Hood took a look. "What's wrong with this trace?" One of the dyes threw off a much weaker signal than the others. "That needs to be a little better for the paper," Hood told Smith.

"Aw shit," Smith thought. He knew Hood was right, but he hadn't wanted to admit it to himself. "I had gone as far as I could with the instrument," Smith said. "I was going to have to jump a level in instrument complexity to get better data. It meant another six months, and another rat's nest of technical issues." Again, Hood pushed Smith to keep going.

Hood pushed himself, too, to get the funding. The usual sources, federal agencies such as the National Institutes of Health and the National Science Foundation, weren't interested, largely because of the anti-technology biases within biology at the time. By 1985, the lab made enough progress that Smith, as lead author, published his first paper on the device in a small but influential technical journal, *Nucleic Acids Research*. Now the federal grants started to come in. They had a "working" semiautomated prototype. Smith spent many more months working on the instrument, and its optical capabilities, to try to get a better signal from the dye attached to that fourth unit of DNA.

Then came the moment Hood had been eagerly awaiting. After years of effort, in June 1986, *Nature* published his lab's paper describing the first automated DNA sequencer. That fact alone was testimony to Hood's clout, as the top-tier research publication rarely published papers about instruments. The paper listed Smith as the first author, meaning he was the one who did the most work, and Hood as the last author, meaning he oversaw the lab where the work occurred. The paper's abstract summed up the achievement in clear, succinct, subdued language:

> We have developed a method for the partial automation of DNA sequence analysis. Fluorescence detection of the DNA fragments is accomplished by means of a fluorophore covalently attached to the oligonucleotide primer used in enzymatic DNA sequence analysis. A different coloured fluorophore is used for each of the reactions specific for the bases A, C, G and T. The reaction mixtures are combined and co-electrophoresed down a single polyacrylamide gel tube, the separated fluorescent bands of DNA are detected near the bottom of the tube, and the sequence information is acquired directly by computer.

This was a rare moment for Caltech: a technological tour de force worth publicizing in a big way. Various other supporters, including latecomers such as the National Science Foundation, took a victory lap for their contribution to "gene machines." There was talk of holding a press conference in Washington, DC, but ultimately Caltech invited the media to Pasadena.

The day of the press conference, June 11, 1986, Hood was the only scientist in front of the cameras. Naturally, the public saw him as the inventor of the automated DNA sequencer. The other members of the team—nine coauthors were named on the big June 1986 splash in *Nature*—were anonymous to the public. While the *Nature* article was aimed at scientific peers, Hood always had multiple audiences in mind: media, businesspeople, wealthy donors. Some scientists may have shrugged off the press conference as mere fluff. Not Hood. Mainstream media, especially during the 1980s, had far-reaching influence. The press conference mattered.

On that day, Hood's name was already big in science. Many of the reporters knew it, and they were quick to give the spotlight to the friendly and eloquent voice of science. Hood's singular talent for conveying the power and wonder of science, while connecting it to meaningful real-world improvements in medicine, was on full display. Manual DNA sequencing methods at the time were slow, and they cost one dollar to five dollars per base unit of DNA. Automated DNA sequencing would cost just pennies per base. A world of possibilities suddenly became realistic. As Caltech said at the time:

> This machine will automate many of the tasks
> involved in DNA sequencing now done laboriously
> by hand. Thus, it will allow researchers to analyze
> the structure of DNA molecules far more quickly
> and less expensively than ever before.
>
> Such analysis is vital to understanding genetic
> diseases and cancer, because often only subtle

differences in DNA structure exist between normal
DNA and DNA involved in pathogenic conditions.
To understand these diseases, researchers must
discover what these subtle differences are and how
they cause an illness. ...

Just as the revolution in physics occurred some
20 years ago with the development of accelerators
which could smash atoms, so has the revolution in
biology occurred with machines which allow us to
analyze and synthesize proteins and genes.

This big science is fundamentally different from
the big science of physics, in that these instruments
are far more modest in cost, perhaps $300,000 to
$400,000, thus permitting each research institute,
genetic engineering company, and university,
to set up central facilities which will allow their
researchers to have access to this very powerful
technology.

The next morning's *Los Angeles Times* touted the achievement:
"Caltech Scientists Develop Super-Fast DNA Analyzer." The story
made clear that this was history in the making.

In a development expected to further accelerate
the already rapid pace of research involving genetic
engineering, scientists at Caltech announced
Wednesday that they have developed an automated
instrument that analyzes the chemical components
of DNA hundreds of times faster than was previously
possible.

The new machine, which "reads" the chemical
sequence of DNA within genes, can do in one day
almost the same amount of work now performed
by a skilled technician in one year. Scientists said
its development could have far-reaching effects on
biotechnology and on research into genetic disease
and cancer.

Biology would leap "from a cottage industry into the big time,
with the ability to solve large-scale problems," James Brown, director
of molecular biophysics at the National Science Foundation, told
the *Los Angeles Times*. He added that every major lab would need at
least one of the machines. Hood was euphoric. Scientists, he told
the newspaper, will "learn as much in the next 15 years as we have
learned in the previous 2,000." It was a line he repeated many times.

But DNA sequencing hadn't been automated, exactly. Not
entirely.

George Church, then a graduate student working under the legendary Walter Gilbert at Harvard, said the *Nature* paper was a bit
overblown. "On inspection, I could not find any automation," Church
said years later. "The reactions and gels were done manually, the data
came like chart recordings, and these were interpreted by pencil on
the charts." The original prototype described in the *Nature* paper ran
the gel through a capillary tube, which in those days couldn't allow
samples to pass through fast enough for a truly automated machine.
Other biologists found the grandiosity of Hood's descriptive language
off-putting. Like Church, they knew that the prototype instrument
was still just that—a prototype. It could deliver the sequence of only
a few hundred bases of DNA in a single run. That sometimes wasn't
enough to determine the order of a single gene.

At the June 1986 press conference, Hood didn't dwell on the
details or the technical shortcomings. He also didn't single out any

of his deputies—Lloyd Smith, Mike Hunkapiller, Tim Hunkapiller, or others—to praise their contributions. Smith recalls:

> While I was in Europe, I asked him "Should I come back for the press conference?" He said "No, no, just enjoy your trip. I'll represent you." Then I was told by everybody when I got back that Lee never mentioned my name in the press conference, despite the fact it was all my work and I'd been doing it for five years. That was upsetting.

Smith wasn't the only one upset by Hood's performance for the cameras. Some were so angry they never spoke to him again. Tim Hunkapiller said he and other members of the lab confronted Hood.

> We had a meeting with Lee … where we were so mad, ABI was mad. They had to get Lee to agree to show a slide from then on. We broke down who did what. I made him the slide, and he had to agree to show it.

Neither Tim Hunkapiller nor Lloyd Smith recalls Hood apologizing or showing any remorse for omitting their names at the press conference.

The official news release from Caltech was more careful to give credit to other members of the team: "[T]he machine uses a laser, colored dyes, and a microcomputer to automatically analyze the structure of DNA molecules. Dr. Lloyd Smith, working in a research group headed by Dr. Leroy E. Hood, developed the concept, and with Mrs. Jane Sanders and Dr. Robert Kaiser, originated a new chemistry and constructed the first instrument for automated DNA sequencing."

For sure, Hood wasn't the first leading man of science to take excessive credit for a team accomplishment. He won't be the last. That sort of selfish behavior is part of a long tradition in science. Hood certainly remembered his own adviser, Bill Dreyer, stepping on him as a graduate student. When Smith left the Hood lab for a faculty position in 1987, it was with mixed feelings about his boss. On one hand, Smith felt fortunate to be in Hood's lab, with the right skills at the right time in the right place to tackle such an important project. On the other hand, it was hard to watch Hood get credit when Smith had done so much innovative and technical work. The two men weren't exactly friends, but on occasion, Smith invited Hood to stay at his house when visiting him at the University of Wisconsin. Years later, Smith reflected on his adviser:

> He's never sabotaged me. There are PI [principal investigators] out there who will sabotage their people. Not Lee. He's either been neutral or supportive of me. ... But Lee is very egocentric. He can't help it; it's just the way he is. When he didn't mention me at the press conference, it wasn't that he was being vindictive. He was just wrapped up in the moment, and it was all about Lee, the quarterback and all that. ... There's not that much loyalty you'll find among his past people. He's not the kind of person who engenders loyalty, because of this behavior.

So Caltech and Hood took home the accolades, while Applied Biosystems did the yeoman's work to make machines that were truly good enough to commercialize and, eventually, to complete the genome project.

There wasn't exactly an orderly passing of the baton from academia to industry. In 1983, while Lloyd Smith was sweating over the chemistry challenge of attaching tags to DNA in Hood's Caltech lab, Mike Hunkapiller left to join Applied Biosystems (ABI). Hunkapiller was considered a cofounder of ABI, and now it was time for him to step out of Hood's shadow and advance his own career. That year, ABI put a former Hewlett-Packard engineer named Kip Connell to work on its prototype DNA sequencer.

Even though Applied Biosystems was more than 370 miles away from Caltech, Hunkapiller and others at the company stayed in touch and the two groups worked on DNA sequencers separately, but in parallel. Collaboration between the company and the academic lab, which was not unusual for such a project, proved critical. Applied Biosystems used the fluorescent tags that Smith developed, for example. The company also created a proprietary fourth dye that threw off a more distinct signal, effectively tagging all four bases of DNA. ABI didn't share this innovation with its academic partner at first. ABI also saw the capillary-tube bottleneck in the Caltech system and realized that it wasn't practical for a commercial machine. The company shifted gears and went to work putting the tagged DNA through a "slab-gel" system, which could handle samples at higher throughput. "ABI had abandoned the tube gel system long before the *Nature* publication in favor of the slab-gel-based design," Mike Hunkapiller said.

Scant mention was made of Applied Biosystems' role in the development of the machines at the victorious Caltech press conference. Behind the scenes, the company continued to bang away at technical issues. Seven months later, in early 1987, the ABI 370 sequencer was fast and reliable enough to be delivered to its first commercial customer. ABI continued to incorporate newer and better technologies. Eventually, ABI's automated DNA sequencers were used to complete the Human Genome Project. The big leap came a decade later when scientists ditched the still-cumbersome slab-gel sample loading method in favor of capillary tubes, which had been improved since the Caltech team favored them way back in the mid-1980s.

Who invented which parts of the DNA sequencer? So many people were involved, with so many different versions of events, a federal judge would later have to sort through it all. Mike Hunkapiller, in his telling of the story a few years later to author Robert Cook-Deegan, went out of his way to say that ABI engineer Kip Connell's contributions had been downplayed as well—that Connell deserved to be named as a coauthor on the seminal June 1986 paper in *Nature*. Hunkapiller wrote to author Robert Cook-Deegan in 1991:

> While Lloyd clearly played a major role in the development efforts at Caltech, he consulted with, but did not play a major role in, those at ABI, which developed along an only partially overlapping path.

> Moreover, in any effort as complex as automating DNA sequencing, descriptions of any one person "driving most of the road" by themselves are false and misleading. As an example, although Lloyd's efforts to find the appropriate dye set and attachment methods were pioneering, they were never commercially implemented. The mobility-altering effects of his dyes prevented accurate automatic base calling, and the chemical linkers he devised were not sufficiently stable for reliable use. Steve Fung and the organic chemistry group at ABI developed four different dye/linker combinations to deal with these problems.

The point was made. The DNA sequencer was a team effort. Hood's grab for credit came with a price. It hurt his relationships at Applied Biosystems, and that created an opening for other scientists. One was J. Craig Venter. Then a little-known scientist at the National Institutes of Health, Venter read the *Nature* publication in June 1986 with interest. He was impressed by the Hood group's "remarkable advance" with the four-color automated sequencer. He decided he wanted the NIH to become the first test site for the Applied Biosystems commercial-grade DNA sequencer, the ABI 370. Over the following months, Venter maneuvered into position, scraping together enough government funds to make his pitch.

To researchers, being the first test site is a coup: it means a crucial competitive advantage over peers. To instrument developers, choosing a beta tester is a critical strategic decision. Companies want to select the best research teams, the most influential teams, the ones with the cleverest research agendas to really show off what the machine can do for all other potential paying customers. Given Hood's years of effort in the field, his lab should have been at the top of the list. It wasn't. His goodwill reservoir was depleted. Hood asked Applied Biosystems to give his lab, for free, a $110,000 machine, which the company had spent years and millions of dollars to develop. The company said no. Hood had overplayed his hand.

Venter was elated to be chosen. When the machine arrived at his lab in February 1987, he wrote in his autobiography, he felt "this was my future in a crate." Working in concert with Applied Biosystems, Venter helped iron out some of the technical bugs with the chemistry in the early machines.

Hood was still a key figure in the genomics community. He continued to meet with leading biologists to talk genome strategy. He maintained a sequencing technology development team in his lab, and he did some DNA sequencing in his immunology research. But behind the scenes, his research was losing momentum. By 1988, two years after the landmark paper in *Nature* and the fateful press conference, the Hood group wasn't even regularly using its own heralded machine. It was still doing DNA sequencing the old-fashioned way, by hand. Hood, without a prototype in his lab to constantly fiddle with and improve, had given up his first-mover advantage in automated DNA sequencing.

There were a number of reasons in the late 1980s for Hood and other scientists to move cautiously with what amounted to a version 1.0 automated DNA sequencing tool. The machines were expensive, and so were the chemical reagents necessary to run them. The machines were slow. They sometimes spat out inconsistent results. The software to crunch the data was primitive. The challenges were ordinary for any new technology. But Hood's team wouldn't be the one to solve them.

Years later, when the Smithsonian Institution asked Hood to send the DNA sequencing prototype to them for historic display, he couldn't oblige. It didn't exist anymore. It had been stripped down for parts.

"Lee's group was not doing much DNA sequencing of any type at that point," Mike Hunkapiller said. "He wanted a free sequencer, which Applied Biosystems was not willing to provide for a group more interested in playing around with their own rudimentary development efforts than in sequencing interesting DNA."

Hood's missed opportunity was Venter's golden moment. Venter went on to run Celera Genomics, brazenly racing the publicly funded Human Genome Project to complete the first full genome. A dozen years later, it was Venter, not Hood, who was invited by President Bill Clinton to a White House press conference to announce a historic first: a "draft" of the full code of life had been completed.

One of Hood's contributions was to speed up innovation. Could the DNA sequencer have been developed without Hood's vision and tenacity in the early years? Yes, but more slowly. A team at DuPont was working on the project at the same time, as was a group of scientists in Japan. The DuPont team, in particular, had significant resources and chemistry expertise. DuPont was able to commercialize its own instrument, as was a small company later swept up in a series of mergers into GE Healthcare. However, Applied Biosystems became the dominant player, through steady technical improvement and clever, focused marketing. It captured about 90 percent of the early market for DNA sequencers. Even more importantly, the Applied Biosystems tools got scientists thinking more seriously about the feasibility of sequencing short stretches of interesting genes. A few dreamed in the late 1980s of possibly sequencing the first entire human genome, and what technologies would be necessary to make it happen. Ultimately, in the next decade, the ABI workhorse machines made the Human Genome Project possible. They were the forerunners of greatly improved "next-generation" sequencers marketed to great effect thirty years later by Illumina. This is what Hood started, by setting the vision and overseeing the invention of the first prototype at Caltech. Yet that was not the whole story. It was sort of like taking the opening kickoff of a football game and returning it to the twenty-yard line. The talented and dedicated engineers at Applied Biosystems, a largely anonymous bunch, did the work of marching the ball the remaining eighty yards downfield for a touchdown.

Hood may have slipped behind in the end, but his scientific peers didn't forget how far ahead he was in the beginning. The automated DNA sequencer, over the years, widely came to be considered the greatest achievement of Hood's career. As the impact of the tools became clear, Hood won the Kyoto Prize, the Lemelson-MIT Prize, the Russ Prize from the National Academy of Engineering, and the National Medal of Science—almost every major award possible for the invention except the Nobel Prize.

Hood sometimes acknowledged collaborators during his acceptance speeches. In his later years, on occasion, he could be more gracious. "I assembled a team including a chemist/laser expert (Lloyd

Smith), an engineer/chemist (Mike Hunkapiller), a biologist with knowledge of computer science (Tim Hunkapiller), and myself, a molecular biologist—and one spring day we had a transforming conversation," Hood wrote in the *Annual Review of Analytical Chemistry*. As the years went by, he gave Smith more of the credit. In 2008, in that *Analytical Chemistry* article, he said Smith "played a central role in solving many of the [DNA sequencing] problems."

But the awards Hood won always had Hood's name alone.

Hood had won the public's mind, too. In fact, Michael Crichton's 1991 blockbuster sci-fi novel *Jurassic Park* talks of "Hood sequencers." In that dystopian book, dinosaurs are brought back to life, with horrific consequences, by scientists using techniques of DNA sequencing and cloning. Hood was a sci-fi fan. But he thought the novel—which led to one of the biggest Hollywood box office successes of all time—was "incredibly hostile to science." In the movie, "much genetic engineering is thoughtless or frivolous, the work is uncontrolled, and it is done in secret, in haste and for profit," Hood told the *Los Angeles Times*. "That's simply not correct."

As automated DNA sequencing took off in the late 1990s and subsequent iterations dramatically improved in the next decade, Hood's legacy was cemented as one of the most important biologists of the century. DNA sequencing was historic, and despite a handful of people's anger over not receiving fair credit for building the machines, the achievement was widely recognized in both biology and technology circles. Michael E. Phelps of UCLA, co-inventor of the positron emission tomography (PET) scanner, summed up Hood's contribution: "No single person has done more to create the genomics era than Leroy Hood."

Like many successes, this spawned jealousies and rivalries, and questions of shortcuts or wrongdoing along the way. Hood was always able to brush off these accusations.

Except for one.

DNA Device's Heredity Scrutinized by U.S.
Scientists insist they invented technique without federal funding,
but paper trail suggests they relied on millions in grants.
Inquiry will determine if there is need for payback.

The expose-style headline ran in May 2000 in the *Los Angeles Times*, the publication that lionized Hood and his achievements more than any other. Henry Huang, the former lieutenant who first worked on automated DNA sequencing, was a plaintiff in a civil lawsuit years later that claimed he, too, was a rightful inventor who had been unfairly left off the lucrative patent. Huang's claims, years after the patents were issued, caught the attention of investigative reporters at the *Los Angeles Times*, and later at *USA Today*. The reports had a skeptical, even sinister, tone. The *Times* suggested that Hood and Caltech had cheated not only Huang but the federal government as well out of their fair share of the royalties.

Hood was known in scientific circles for his unflappable, always upbeat temperament. Former students recall a man who would simply shrug, or even smile, when called a lying son of a bitch to his face. He was famous for never holding grudges. But public charges of wrongdoing? That got under his skin. "Someone out there has a vendetta," Hood told the *Times*.

What newspaper reports failed to mention, or downplayed, was who had initiated the lawsuit. It wasn't Huang. It was a competing instrumentation firm called MJ Research—which had a financial motive in overturning the Caltech patents licensed to Applied Biosystems.

Plaintiffs' lawyers spent years poring over lab notebooks and other internal documents in the case. In 2004, US District Judge Mariana Pfaelzer wrote in a summary judgment that Huang failed to make a convincing case. The judge, after hearing so much testimony, made an astute observation about Huang, which easily could have been applied to all the scientists, each with their own egos. "The invention that he now claims to have conceived has received international acclaim and tremendous financial success," Pfaelzer wrote. "Even honest and well-meaning people might be tempted to amplify their role in creating such an important invention."

The litigation, however, wasn't easily dismissed from the start. It caught the attention of investigators at the US Department of Justice. Had Hood used federal funds to invent the DNA sequencer? If he had, then Caltech might owe the feds millions of dollars in royalties, or discounts on purchases of the instruments used in federal labs. The Justice Department took lengthy depositions from Hood and others. Eventually, it dropped its investigation.

The DNA sequencer was the fourth, and final, piece of Hood's broad technological vision for biology. Hood could now credibly claim he led teams that had done pathbreaking work on a series of instruments that opened up a wide array of new frontiers in molecular biology.

Getting all four instruments working together was like a warp drive booster for biological research. "Biologists who once had to kill 1,000 mice to get enough protein to experiment with now need only one or two," the Seattle Times reported in 1992. "A gene synthesis that once took 20 chemists five years to build can now be done by a technician using part of one day."

This kind of leap naturally inspired an entire generation of biologists to think more ambitiously about the kinds of questions they could ask. No longer did they have to settle for focusing on a single gene at a time, just because that's what the tools allowed. Entire new branches of science sprang up, such as computational biology. When Hood was awarded the National Medal of Science in 2013, President Barack Obama said Hood didn't win for a single technology or discovery. Hood was honored for work that created "several new scientific disciplines that have challenged and transformed the fields of biotechnology, genomics, proteomics, personalized medicine, and science education."

Bill Gates wrote a nomination letter for the National Medal of Science. "I see every day the impact of Lee's contributions to the world of medicine and biology," Gates wrote. Hood's collaborations "resulted in the creation of four instruments which would form the technology underpinnings for contemporary molecular biology. The field of genomics owes its existence to Lee's breakthrough work in DNA sequencing and assembly.

"He has made signal contributions to the Human Genome Project and remains a world leader in computational biology. In the field of vaccines alone, these contributions will bear fruit in new therapies, vaccines, and millions of lives saved," Gates wrote.

The DNA sequencer was a financial windfall, too. More than $100 million in royalties were collected in Caltech's coffers over the life of the patents. That's because a licensing agreement with Applied Biosystems gave Caltech a 2 percent royalty on sales. It was the second-most-lucrative invention ever to emerge from Caltech. Of that royalty flow, 85 percent went to Caltech, while the remaining 15 percent was split among the inventors listed on the main patent— Mike Hunkapiller, Tim Hunkapiller, Lloyd Smith, and Lee Hood. Hood discovered early on that if he donated his personal share of the royalties to his lab, he could get significant matching funds from Caltech, which he could plow back into the lab to cover flexible expenses. Long after he left the institute, Hood continued to donate his royalties to Caltech. For a while, Hood tried to persuade his co-inventors to do the same, but there his influence had waned as well.

No matter. Hood's vision was beginning to be realized. He was on top of the world, looking invincible. As usual, that's when things started to come unglued.

Chapter 11

The Clash at Caltech

As technology advanced in the 1980s toward automated DNA sequencing tools, scientists began to imagine—and then debate—truly audacious goals. Long before the Human Genome Project seemed feasible, a few leading scientists wondered whether it would be possible to sequence all six billion units of DNA in the human genome. And whether they should.

"Nonsense" was Hood's first reaction. Lloyd Smith had just published his first paper and had a "working" semiautomated prototype; Kip Connell at ABI was working on the crucial fourth dye. The triumphant 1986 paper in *Nature* didn't exist—and neither did anything close to high-speed sequencing.

The new chancellor of UC Santa Cruz, Robert Sinsheimer, had just gotten a $35 million gift to set up an institute dedicated to studying the human genome. He wanted to do something big with the money. Sinsheimer invited some prominent friends to Santa Cruz for a summit in May 1985. John Sulston of Cambridge University, Robert Waterston of Washington University in St. Louis, David Botstein of MIT, Walter Gilbert of Harvard University, and Leroy Hood of Caltech, among others, were called upon to discuss the possibility of sequencing the human genome.

At first, Hood was among the most skeptical there. By the end, he had turned into an advocate. He was captivated by what business schools call a "Big Hairy Audacious Goal." It's like a moonshot that motivates people to do their very best work, to plow through technical challenges that seem too daunting at first.

Hood recalled his change of heart:

> I went to the [Santa Cruz] meeting with the idea
> that this was nonsense. How would you? But Wally
> Gilbert was a passionate and articulate advocate, and
> I was almost instantly won over. I mean, I think I was
> guilty of exactly what I just said a lot of other people
> are guilty of: that is, a narrow view. I hadn't thought
> about the broader implications, hadn't thought about
> the fact that technology could solve a lot of these
> kinds of problems.

Some scientists thought it would take a hundred years, or maybe even be impossible, to get a complete, accurate human genome. "Their collective conclusion: bold, exciting—but simply not feasible," *Science* magazine reflected on the Santa Cruz meeting.

The machines weren't ready for such a daunting task. A year after the Santa Cruz meeting, Hood's DNA sequencing team published its seminal paper in *Nature*. Later that summer of 1986, Hood was evangelizing about a historic journey-in-the-making. At a summit convened by the Howard Hughes Medical Institute, Hood compared the sequencer to the automobile. The original machines were like Henry Ford's Model A—the first mass-produced car, which had a top speed of about thirty miles an hour. To sequence the genome, Hood said, scientists needed a Ferrari. That was many years away. He wanted to focus on building the Ferrari first. "Hood urged that the emphasis fall on developing new technologies during the early years, so that later efforts [to sequence] would be faster and more efficient," the author Robert Cook-Deegan said.

Hood testified passionately before several congressional committees to get funding for a genome project. At one session, in February 1987, he showed up with handwritten notes sketching out plans for technology development and physical mapping. He advocated some basic DNA sequencing of model organisms, like yeast and bacteria. And—to make the future medical benefits clear—he suggested determining the sequences for targeted regions of the human genome. He asked for $200 million to $300 million a year. Eyebrows went up. That was tenfold higher than some earlier estimates. "Will the budget drive the vision, or will the vision drive the budget?" Hood asked. At a second hearing a month later, Hood again made a forceful argument. On his way out to catch a taxi to the airport, he asked a young Cook-Deegan, then a legislative staffer, "Is that it?"

Puzzled, Cook-Deegan asked what he meant.

"Do we get the money?"

Startled, Cook-Deegan told the prominent scientist that the hearing was just one of many. "It was far from a done deal," Cook-Deegan later wrote in his book, *Gene Wars*. "He was a long way from home."

Within three years of that Santa Cruz summit, by 1988, a consensus for the genome project had formed among biology's elites. Hood was a member of an influential committee chaired by Bruce Alberts of UC San Francisco, which recommended that the Human Genome Project go forward.

Congress still wasn't sure. Some members worried that insurance companies or employers would use genetic information to discriminate against individuals. A few worried that genomics would revive eugenics, the 1930s movement to weed out people deemed genetically inferior, a concept used by Germany's Nazis to justify atrocities.

In a thoughtful and inclusive response, Hood organized a series of guest lectures on the implications of genomics, held at Caltech in 1989 and 1990. The lectures—by scientific elites such as James Watson, Walter Gilbert, and Eric Lander—were later compiled in a series of essays. *The Code of Codes*, edited by Hood and Daniel Kevles, became a widely used introductory college textbook. In it, Kevles wrote about the historical lessons from the eugenics debacle of the 1930s, while Hood outlined a tantalizing vision of what society could learn from the genome. As they wrote in the book's preface:

> The human genome comprises, in its totality, all the
> different genes found in the cells of human beings.
> The Nobel Laureate Walter Gilbert has called it the
> "grail of human genetics," the key to what makes
> us human, what defines our possibilities and limits
> as members of the species of Homo sapiens. ...
> The knowledge will undoubtedly revolutionize
> understanding of human development, including the
> development of normal characteristics, such as organ
> function, and abnormal ones, such as disease. It will
> transform our capacities to predict what we may
> become and, ultimately, it may enable us to enhance
> or predict our genetic fates, medically or otherwise.
>
> Unquestionably, the connotations of power and
> fear associated with the Holy Grail accompany the

genome project, its biological counterpart. The
project itself has raised professional apprehensions
as well as high intellectual expectations.
Undoubtedly, it will affect the way that much of
biology is pursued in the twenty-first century.
Whatever the shape of that effect, the quest for
the biological grail will, sooner or later, achieve its
end, and we believe that it is not too early to begin
thinking about how to control the power so as to
diminish—better yet, abolish—the legitimate social
and scientific fears.

Hood and the other elites rallying around the Human Genome
Project were a minority camp in biology. "Alarm, derision, and out-
right fury" was the reaction of most biologists, according to the *New
York Times*, as the Human Genome Project ramped up in June 1990.

To many opponents, the thing sounded radical and the professional apprehensions were real. For starters, the Human Genome
Project represented a drastic departure from traditional ways of
working. Tradition called for "small science": one investigator
leading a small group of careful, meticulous grad students and postdocs. They'd devise an experiment based on a creative hypothesis,
gather data, analyze it, form a conclusion, publish the results, and
repeat. "Small science" was decentralized. Research proposals had
to pass peer review before getting funded, a system designed to
weed out crackpot science. The notion was, and often still is, that
no single place has a monopoly on knowledge. That the best ideas
sometimes sprout up where you least expect them. If you wanted
to discover something like the genetic abnormality underpinning
cystic fibrosis, for example, it was best to have a dozen small competing teams pursuing independent ideas. They would all charge
ahead, day and night, in a quest for glory that would come one day

to the winner, in the form of a trophy publication in the pages of *Science* or *Nature*. Some would call it the "let a thousand flowers bloom" approach to science.

The Human Genome Project was the opposite. It was "big science": practiced in big, central facilities, with big budgets and big teams of people from multiple disciplines of science and engineering. As in a big corporation, people would be organized around pursuing a single overarching goal. The initiative was sure to siphon away much of the federal research budget, which many scientists feared would doom "small science."

Just as bad, in some scientists' view, the Human Genome Project was an assault on the traditional scientific method. They weren't comfortable with the idea of first generating massive reams of data and then working backward to look for hypotheses. Anyway, what was the point in sequencing the complete genome when 95 percent of the code appeared to have no biological function? Why go through all the trouble of gathering this massive trove of data if nobody knew what to do with it, how to interpret it? Didn't it make more sense to start with proteins of known function, like insulin, and work backward to get the genetic codes? The genome project struck many as dull, industrial-like, crank-turning work that would yield useless "junk DNA."

Elite universities like Caltech weren't places for crank turning. They thought of themselves as places for creative geniuses—individual, quirky, and fiercely independent geniuses. Hood and his smooth-talking, techno-driven, interdisciplinary team-oriented vision of biology had been tolerated in small doses. The microchemical facility, running the four automated instruments, was one thing. The idea of a much bigger site for sequencing the entire human genome? Bringing in whole teams of people who didn't, frankly, belong at Caltech?

That was going too far.

Seymour Benzer, a Caltech geneticist from an older generation, voiced the sentiment in an oral history:

> Interviewer: I've heard Lee Hood express the opinion that if the current and future generation of biologists doesn't come to grips with the fact that the science is becoming increasingly computerized and quantitative, they're going to fall by the wayside. Do you agree with that?
>
> Benzer: Well, that's Lee Hood's idea, and he's riding with the tide. I don't think the history of creative thought has been built by people who were riding with the tide. It's been built by freaks who were using their brains. So I don't like it. As Max Delbrück used to say, the universities are a haven for freaks. Delbrück was a freak; Einstein was a freak; Kurt Gödel was a freak. I think there's some kind of natural evolution in a science when it becomes too successful.

Power brokers like Watson, Gilbert, and Hood dug in their heels, fiercely advocating the project. They argued that it didn't make sense for decentralized teams to sequence the same gene; a centrally organized effort was more efficient. They argued that all of biomedical research would benefit, because the genome project would speed up the search for genes that cause all kinds of diseases. The project also would force improvements in technologies, which would spin off in interesting new directions.

Their rhetoric was grandiose and often condescending. Gilbert frequently used religious language. "The total human sequence is the grail of human genetics ... an incomparable tool for the investigation of every aspect of human function." Watson told the *New York Times*, "It's essentially immoral not to get it done as fast as possible."

Money wouldn't be siphoned away from other worthy projects, they argued. The funding represented a new line-item appropriation from Congress, and it equaled only a small percentage of the total budget of the National Institutes of Health.

But the fact was that government funding had already been curtailed for many young scientists in the late '80s. Some thought that the big guys like Watson, Gilbert, and Hood—who already had plenty of funding—were lining up against the anonymous little guys. "I had a brilliant young scientist say to me recently, 'The fat cats are all getting the cream, while I'm sitting here starving,'" Martin Rechsteiner, a biochemist at the University of Utah, told the *Times*. James Wyngaarden, NIH director during the early days of the genome project, was unmoved. He dismissed critics as "journeymen biochemists who may be having a hard time competing."

The Human Genome Project fell behind schedule almost from the start. Critics saw that as proof that the project was more challenging than Congress had been led to believe. And they knew any great advances in treating cancer and other diseases would be many years away. The whole thing seemed a bit too slick, packaged into the media-friendly sound bites that Hood was so good at. Rechsteiner spoke for many critics when he said, "The Human Genome Project is bad science, it's unthought-out science, it's hyped science." Harvard geneticist Richard Lewontin argued that the answers of life aren't all found in the genetic code, and therefore the project's benefits were being exaggerated. Social and environmental factors—confounding and difficult to measure—played a huge role in life. Much of this nuance was missing from the public debate, he said, which made it sound as if genes were all that mattered. Lewontin acidly observed the "evangelical enthusiasm of the modern Knights Templar and the innocence of the journalistic acolytes whom they have catechized."

Some critics went so far as to write letters to Congress, urging lawmakers to pull the plug.

Sequencing of the genome was indeed technically challenging. Biologists had long debates about how to achieve the best combination of speed and accuracy. There was no machine, then or now, that could just process a biological sample and determine the exact order of all six billion units of DNA in a straight line from start to finish. Cells needed to be shredded into millions of fragments, so that the machine could read short stretches of DNA at a time. Each stretch then needed to be carefully aligned and mapped to its proper location on a chromosome. It was like sketching a map of the world, one town at a time. Region by region, researchers would build a high-resolution, comprehensive, global map.

The classical mapping process was tedious, and the early semi-automated technology for reading short sequences of DNA was slow and error prone. Even enthusiasts for the genome project, like Hood, knew it would take years before the technology could handle high volumes of samples at high speeds. That's when the project could shift into high gear and scientists could seriously contemplate assembling a whole genome. With the right mix of technological improvements, Hood argued, the project could be completed in fifteen years, not one hundred. Skeptics scoffed at what they saw as wild optimism.

But the advocates prevailed. Congress authorized the money. The Human Genome Project officially began on October 1, 1990.

Hood was on top of the world.

By his early fifties, he had a string of achievements in immunology and technology to his name. He was a proven fund-raiser, fetching tens of millions in federal grants, donations, and corporate sponsorships. He packed lecture halls with spellbinding talks. He was a go-to source on biology for the media. Job offers came regularly from universities and corporations alike. Harvard asked Hood

Leroy, Doral, and Myron Hood in Montana in the mid-1940s.

Siblings Doral, Myron, Leroy (back row) and Glen (in front), in 1949.

ABOVE The quarterback.

RIGHT Valerie and Leroy met at a theater competition.

Valerie and Lee Hood (far left and far right) with his parents, Myrtle and Thomas (center right), in the early 1970s.

THE DAILY

OF THE UNIVERSITY OF WASHINGTON

INDEX
- COMICS .... PAGE 14
- EDITORIALS .... PAGE 6
- CALENDAR .... PAGE 11
- SPORTS .... PAGE 21

VOL. 99 NO. 6

gives record $12 million

$12 million gift: Leroy Hood, Bill Gates and UW President William Gerberding at a press conference yesterday. Microsoft founder Gates donated the money to the UW for the formation of a molecular biotechnology department. Hood is to chair the new department.

Kerry Dahlen

Hood joined the Caltech faculty in 1970.

TOP LEFT Bill Gates' first major act of philanthropy brought Hood from Caltech to the University of Washington in October 1991. UW President Bill Gerberding held a press conference with the two men.

BOTTOM LEFT Gates and Hood stayed in touch after Hood left UW.

Phil Leder (far left), Susumu Tonagawa (left), and Lee Hood (far right) share the Lasker Award in 1987. Mary Lasker is in the center.

Start-up days: Lee Hood, Alan Aderem, and Ruedi Aebersold cofounded the Institute for Systems Biology in Seattle in 2000.

Valerie and Lee Hood traveled with Bill Gates and his future wife, Melinda French, on their first trip to Africa, in 1993.

High-school debate taught Hood to make eye contact and speak without notes. Throughout his career, wherever he gave talks, he made a point of seeking out graduate students he could recruit.

Family time: Hood was deeply proud of both of his children and became closer to them in later years.

RIGHT While at Caltech in the 1970s and 1980s, Lee and Valerie spent summers with the kids, Marqui and Eran, in Aspen, Colorado.

RIGHT Hood often sought peace and inspiration in the mountains.

BELOW Valerie Logan and Leroy Hood in later years, as Valerie declined into Alzheimer's.

to apply for its presidency. Roche, the enormous health-care company, offered to increase his salary sevenfold. Monsanto beckoned with another lucrative bid for the Caltech star.

Always, the answer was no. Climbing the administrative ladder meant moving further away from science. Industry, in his view, was much too fixated on short-term, incremental advances intended to boost profits. It wasn't the place for a visionary. He was looking out ten, twenty, fifty years into the future. For that, nothing beat being Caltech's chairman of biology. He was doing great science, at a great place, surrounded by great people. Hood loved it.

But not everybody loved him. As the human-genome debate raged nationally, a similar dispute played out on a smaller scale at Caltech. The object of alarm, derision and outright fury, in this case, was Lee Hood himself.

It wasn't just that Hood's vision for big science in biology was increasingly out of sync with the small science favored at Caltech. It was that Hood's desire to do big science in his own lab required elbowing out his colleagues for money, space, and equipment. That didn't sit well with the elbowed. The Caltech biology faculty, who had asked Hood to be chairman of their department—twice—began to turn on him.

Melvin Simon, whom Hood recruited to the faculty, recalled the tensions in a Caltech oral history interview with Shirley K. Cohen:

> Simon: Lee still got what he wanted and needed; he got plenty of space at the Beckman Institute, and it was very productive. But it was getting egregious, more and more.

> Cohen: You mean, Lee's taking over more and more of biology?

Simon: Everything! Lee had a lab at the bottom of
the Kerckhoff [building]. Lee had the top floor of
the Braun building. Lee had a whole wing of the
Beckman Institute and 120 people he really couldn't
possibly have kept track of. I think a number of
people realized that it just couldn't go on. As I recall,
there was some sort of petition to the provost that
something had to be done.

Something was done.

In a coup led by the Caltech biology faculty, Hood was ousted as
chairman. He never saw it coming.

If he had been a different kind of man, Hood might have noticed
that he was increasingly upsetting his colleagues. For example:

He and another scientist, Eric Davidson, scored a hefty (for
the late 1980s) $12 million grant, spread over five years, from the
Lucile P. Markey Charitable Trust. With the money, Hood said, he
created facilities and resources for the entire department to share.
Ellen Rothenberg, a faculty member then and now, put it differently.
"When the facilities group was establishing a shared DNA synthesis
facility, and a peptide synthesis facility, it was obvious how useful
they were to Lee's work," Rothenberg said. "It wasn't so obvious how
useful it was to other people's work. Or how cost-effective it might
be compared to outside vendors." Neurobiologists and other special-
ists thought Hood already had enough toys. It was their turn to get
some, too.

As chairman, Hood called faculty meetings on short notice,
whenever he happened to be in town. Members of the division
often couldn't attend and have their voices heard. Davidson, Hood's
closest ally, was functioning as a deputy chair—a powerful post,
given that the chairman was often away on travels. Davidson didn't
see disenfranchisement, though, only jealousy of Hood's growing

fame. "They felt they were excluded from the governorship, the division wasn't democratically run, that Lee put his objectives ahead of other people's objectives," Davidson said. The faculty complaints, Davidson said, "were complete bullshit."

The Hood lab was growing at a startling rate. During just one year, 1988, the lab reported that it worked on an astonishing seventy-four research projects. The next biggest group, led by Davidson, had twenty-three projects. Most labs had a dozen or fewer. Hood also ignored the chaos that such a big lab sent rippling throughout the department. As Lloyd Smith had discovered, it was common for graduate students and postdocs to show up in the Hood lab on their first day with no one realizing they had been hired and with no physical space for them to work. "People in the department found members of Lee's lab stationed in every open broom closet," Rothenberg said.

It didn't help that Hood found it hard to turn away young, job-seeking scientists who shared his enthusiasm and stroked his ego. Not all of them belonged. "Lee has a tendency to pick up strays," Roger Perlmutter said.

The quality of the Hood lab, which had been first-rate in the late '70s and early '80s, began to decline. "A lot of people who came out of Lee's lab were very successful," said Joan Goverman, a Hood postdoc at Caltech from 1983 to 1992, who later followed him to the University of Washington. "But were we scientifically mediocre, if you crunch numbers on a per capita, per dollar, per publication basis? Yeah, we were mediocre."

By the end of 1988, the Caltech biology faculty was in a state of mutiny. Hood and Davidson were isolated in their own clique. The neurobiologists and the rest of the "small science" camp in molecular biology were aligned in opposition. Their petition landed on the president's desk: remove Lee Hood as chairman.

Caltech's president and provost had to make a tough decision. They viewed Hood as an important asset to the university. He had a track record of excellent science, recruited good people, raised a lot of money, and generated glowing press for Caltech. He'd gotten the Braun building built and filled it with young, top-notch faculty. Disputes over his use of "unrestricted" funds, the Applied Biosystems licensing deal, and the Arnold Beckman imbroglio felt like distant memories. He was that rarest of individuals—a first-rate biologist and first-rate popularizer of biology. "I was amazed at how well he could present biology information to a lay audience," President Thomas Everhart said.

But Hood frustrated them, too. He didn't hide his disdain for "bureaucracy." It was among the dirtiest words in his vocabulary. Bureaucrats, in his mind, were people who didn't understand science, didn't understand his brilliance. At best, they could support his science (and, of course, he leaned on them every day), but too often, they got in the way. They were to be avoided whenever possible. Administration never moved fast enough for him.

So he did things his own way. For example, it wasn't unusual for Hood to dash across town to give a speech to some foundation or civic group and ask for money on the spot. Once, Mike Hunkapiller recalled, Hood spoke to a women's group that agreed to give him $250,000 one morning. That afternoon, he ran back to the office and asked Hunkapiller to write a one-paragraph summary of the work, then fax it right away. The check arrived within a week: instant gratification! It was exactly how Hood thought fund-raising ought to go. But that's not how fund-raising worked at Caltech, or any other university. There were proper channels. Messages needed to be coordinated when approaching donors—especially the most powerful ones—so as not to duplicate efforts or send mixed messages.

An undated, handwritten note from Caltech President Murph Goldberger to the provost described Hood's Lone Ranger tactics: "infuriating."

215

But the festering, cumulative resentments were no longer Goldberger's problem. He had stepped down in 1987. The institute hired Thomas Everhart, chancellor of the University of Illinois at Urbana-Champaign. Everhart was an electrical engineer by training, known for his work on electron microscopes. He was a member of the prestigious National Academy of Engineering. But he didn't automatically have the respect of everyone at Caltech. Many scientists—even at an institute devoted to both high-minded scientific exploration and technological application—looked down their noses on practical-minded engineers. Everhart was careful not to rock the boat. He struck a pose as a modest leader. "You ask a lot of questions and you listen," Everhart told the campus newspaper. "You don't come in with a lot of preconceived ideas."

So when complaints started rolling in about Hood, Everhart listened. He could see both the jealousy and the chaos. He also saw an empire builder. At Caltech, Everhart said, no one individual should be bigger the institution. When the faculty signed a petition to oust Hood as chairman of biology, Everhart did not halt the rebellion.

A couple days before Christmas 1988, Provost Barclay Kamb called Hood into his office. He started with the good news. Hood had just won a big grant from the National Science Foundation to support an integrated nucleic acid and protein research facility at Caltech. It was one of only a few such centers of excellence around the country. The grant had unusual provisions that supported public outreach and science education, something Hood cared deeply about. It was a validation of Hood's visionary work on the fully integrated microchemical facility. Kamb congratulated him.

Then came the bad news. Kamb said there was a rule that the administrator of an NSF-funded center couldn't simultaneously have other administrative responsibilities. "You've got your hands full, you've got a new thing," Hood recalls Kamb saying. "I want you to step down." With two years left on his term as chairman of biology, Hood was being fired. He would have a few weeks to wrap up his business as chairman. He would no longer be part of the powerful group of seven division chairs who set policy and who met regularly with the board of trustees. But he could continue as a faculty member working on his vision.

Everhart hoped it was a compromise that would allow Hood to coexist with his colleagues. Hood protested. He knew that other scientific administrators had found ways around such restrictions. "I don't need to step down," he argued. But the decision had been made. The department had lost confidence in the chairman. "I think it would be better if you do step down," Kamb told Hood.

Benzer, the Caltech geneticist, was among those who sought the change. "Lee Hood was asked to step down before he wanted to," he noted in his Caltech oral history. "He already had so many enterprises that kept him away from Caltech 90 percent of the time, if not more ... I think it was the right time for him to step down."

Twenty-five years later, Davidson recalled the ouster with bitterness:

> It was a secret cabal. What was surprising was that
> the administration, led by a weak president, and
> an equally weak provost, caved. They should have
> said, 'We'll look into it.' They didn't do any of that.
> ... There was no due process, not even a serious
> accusation of wrongdoing.

Nothing was said publicly for a few weeks. The February 1989 issue of *Caltech News* said Hood "resigned" as chairman to devote his full attention to the new NSF-funded center. The Caltech annual report said he resigned "to devote his energies to his large research group."

In public, Hood continued to project his upbeat persona. His cheerleading for the Human Genome Project was relentless. But in private, he was deeply hurt and humiliated. He had Caltech in his blood. He was a distinguished undergraduate, graduate student, faculty member, and biology chairman. "After everything I had done, I saw that as a really ungrateful and ungracious way to do things," Hood said twenty-five years later.

Looking back, Everhart said the dismissal could have been handled more delicately. Kamb "was not as gracious as he could have been," Everhart said.

Hood had no idea things were about to get much more miserable.

The new chairman of biology, John Abelson, sought to "overturn everything Hood had done," Davidson said. Much like the year after an election in Washington, DC, when a new president brings in his own people and sweeps out those loyal to the previous administration, Davidson said he and Hood were "ostracized." They were kicked off important decision-making committees. An internal investigation probed whether Hood and Davidson misappropriated some of the $12 million they had collected from the Markey Trust.

"Anybody who was close to Lee was persecuted, basically," Davidson said. "Nobody who was close to Lee was put on any committee of significance for about ten years. There was a real vendetta here."

Life under the new regime was shocking for a man used to getting his own way. "Everything Lee tried to do was blocked," Rothenberg said. "Space was taken away, positions were taken away. They were basically trying to hound him into leaving." Hood was careful never to let anyone see him sweat. He complained only to Davidson behind closed doors. "The friction was really beginning to get to me," Hood said.

For the first time in his career, Hood started to think seriously about leaving his intellectual Eden. He still had the energy of a man twenty years younger. He thought his best work was ahead of him. Maybe he would have to do that work elsewhere.

Discreetly, he inquired at UCLA. Hood had a friend there, Michael E. Phelps, who shared Hood's unified vision for technology and biology. Phelps jumped on the opportunity and tried to line up funds from a prominent Los Angeles philanthropist named Norton Simon. "You do not fool around," Hood wrote in a letter to Phelps in May 1990.

The dean of the UCLA School of Medicine, Kenneth Shine, moved to strike a deal. He dangled the possibility of Hood getting a federally funded genome sequencing center on campus, with extra support from the prestigious Howard Hughes Medical Institute. Hood asked UCLA to take fifteen of his deputies, provide the space he desired, and buy $1.25 million in new equipment. He wanted some of the administrative portion of his federal grant money to be set aside as "discretionary income" for his lab.

But Hood's interest fizzled. Ultimately, Hood said he doubted UCLA had the space and resources.

Hood chose to stay at Caltech and deal with the misery. But word of his restlessness got back to President Everhart. Everhart had a careful balancing act to consider. By doing nothing, he risked losing one of the institute's biggest stars. Yet if he bent over backward to keep Hood, he risked further antagonizing the biology faculty.

"Move heaven and earth" to keep the star biologist, urged science historian Daniel Kevles, a Caltech faculty member from 1964 to 2001. He and others outside the biology faculty saw an opportunity to dominate the age of genomics with one of the leaders in the emerging field of the 1990s—even if it meant setting up a freestanding institute affiliated with Caltech. "It seemed to me Lee was doing really good, important work. He was crossing the divide between biology and medicine, he was doing it with important innovations and instruments. You can't run that kind of operation with five or six people," Kevles said. "If the biology operation couldn't accommodate Lee, then he would have to be the pearl in the oyster. You ought to find a way to house the pearl off-site, out of the division, if you have to."

Everhart got up at 4:30 or 5:00 a.m. to go running at the Caltech track, where he knew he could find Hood alone. Hood was a fast runner. Ego still bruised, Hood didn't slow down to the president's jogging pace to listen. Everhart had to huff and puff to keep up as he made his proposal:

Hood could set up a quasi-independent cross-disciplinary institute. Hood's institute could operate like a smaller version of the successful NASA-backed Jet Propulsion Laboratory, which maintained a connection to Caltech. The new institute would allow Hood the flexibility to raise his own money, manage his space, and pursue his own research agenda. He could maintain access to great students and faculty at Caltech.

Hood listened, but he didn't like what he heard. His institute's facility would have been in La Cañada, ten miles away from campus. "I didn't think I could recruit good people there," Hood said later. He also felt "a bit cynical" that the promises would fully materialize. He asked for control over all personnel decisions—including hiring and firing—and was denied. "The way faculty members are chosen has a lot of checks and balances, and it's one of the ways an institution gets to be good," Everhart said. "I wasn't willing to change that for anybody, including Lee Hood."

Plus, Caltech was too small to become one of the new sequencing centers for the Human Genome Project. Hood imagined himself playing an important role in the project, perhaps as the manager of a major sequencing center. Hood was among a handful of eminent scientists who finessed the project through the Department of Energy when it ran into roadblocks at the National Institutes of Health—a bastion of traditional "small science" biology.

The Department of Energy was an unlikely ally, but in a way, it made sense.

By the end of the Cold War, Congress was considering cutbacks to the big national nuclear research labs like Los Alamos and Argonne. One aspect of the research, however, was biological—especially links between radiation exposure and cancer. The genome, a big project that required big scientific resources, was in some ways a natural fit for federal labs that needed some work to do. Hood made fast friends at the Department of Energy who were naturally interested in preserving their own budgets. Hood sang the praises of the Human Genome Project on Capitol Hill throughout the late 1980s. After the decision in 1990 to proceed, the genome project was slated to get $200 million a year in federal funding. The bounty would be divvied up among a handful of sequencing centers. Other universities were better positioned than little Caltech to get a piece of the action.

As the UCLA opportunity faded away, Hood turned to the University of California at Berkeley. Not only was it a site of excellent chemistry and biological research, it was also next door to the Department of Energy–funded Lawrence Berkeley National Laboratory. That meant he could keep one foot in academia and one foot in heavy-duty genome sequencing work—and recruit in both places. Hood spent a whirlwind day in September 1990 touring Berkeley facilities. He was invited by Charles Shank, the director of the Lawrence Berkeley National Laboratory. Shank gave him rockstar treatment. By Thanksgiving, Shank and Hood had agreed on the outlines of a deal. "Before getting into the details, let me simply say how delighted I am about the accord we have reached," Shank wrote to Hood on November 30.

The details: Hood would be director of the Lawrence Berkeley Human Genome Center. He would get a total of forty-five thousand square feet in several buildings—equal to about half of the entire

Braun building he had raised money for at Caltech. Many other items on Hood's wish list were granted. "You will be in charge of Center operations," Shank promised. "Steps are already under way" to recruit junior colleagues whom Hood wanted to bring from Caltech. The director of cell and molecular biology agreed to "sign off on any grant proposals that you recommend." Hood was promised a faculty appointment in both genetics and chemistry.

But when the total budget came in writing, it was $2 million to $3 million short of what had been discussed verbally. When Hood questioned the discrepancy, Shank told him he should just sign. They could work out the additional financing later. Hood had doubts.

Antsy as he was to leave Caltech, Hood wasn't desperate enough to accept a vague assurance. But what happened next was a deal breaker.

The Berkeley biology faculty took a vote. How many supported Hood's new initiative? Two-thirds were in favor, one-third opposed. The vote was alarming. More than anything, Hood craved the respect and admiration of his peers. He wanted a fresh start. Raw from the rejection of his Caltech peers, the Berkeley vote was salt in his wounds. Hood turned down Berkeley—without any real explanation at the time. "This was the most agonizing decision of my life," Hood wrote to Gerald Rubin, a molecular biologist at UC Berkeley, around Christmas 1990. "In the end, the offer was fantastic. I do not think … you could have done any more in recruiting. The decision ended up hinging on many small things, some even difficult to articulate." Years later, Hood explained. "I didn't want to go someplace where one-third of the people were in the same boat as many of the people at Caltech in biology. That didn't seem like an attractive alternative to Caltech. It was really the decisive thing."

Hood wasn't the only one agonizing. Any move he made would affect at least fifty other young scientists. The rumors of Hood's planned move from Caltech and the uncertainty it created had contributed to a lab slowdown. "Everyone in the lab breathed an enormous sigh of relief" when Hood decided to stay, said Lee Rowen, a Hood postdoc at the time.

And then the lab turned upside down. "One of the worst years of my life," Hood said of 1990.

Vipin Kumar was a rising star in the Hood lab. Coming from a postdoctoral position at Harvard, Kumar had gravitated to a hot area of immunology—one that scientists hoped would reveal the underpinnings of multiple sclerosis. Kumar's experiments were getting astounding results. Too astounding, some in the lab suspected. But Hood chalked up those complaints to jealousy.

"Vipin has had an extremely productive stay in my laboratory," Hood wrote in a letter of recommendation for Kumar on April 16, 1990. "Vipin has the energy, drive, and vision to easily become an independently funded investigator. Indeed, I would expect him to rather quickly become a real leader in his field."

The University of Texas was considering Kumar for a faculty job. Hood noted that they might hear some complaints, but that he thought they were baseless. Kumar "is very smart and aggressive. He is also somewhat in a hurry. These characteristics can rub those on a slower track in an inappropriate manner," he wrote. "However, Vipin is congenial and effective in interacting with the technicians he directly supervises. His aggressiveness and impatience is characteristic of many excellent young scientists that I have known. I would put Vipin in the top 10 percent of scientists that I have trained in the past five years."

Aggressiveness and impatience? Hood could very well have been writing about himself as a graduate student. These were qualities that he thought gave him an advantage in the fast-moving, competitive world of science.

But little more than a month later, Hood heard a complaint that he couldn't ignore. A senior researcher in the group, Dennis Zaller, had tried to build on Kumar's earlier published findings. He couldn't. As he tried to figure out why, he showed Kumar's paper—published in December 1989 in the *Journal of Experimental Medicine*—to colleague Mike Nishimura. That evening, May 29, 1990, Nishimura got a sinking, nauseated feeling in his stomach. Once you knew where to look, it was obvious. A key figure was falsified. The next morning, he told Hood.

First, Hood asked Nishimura and Zaller to run some tests. Within a day, the worst fears were confirmed. Hood called Caltech's attorney. He then confronted Kumar, asking him to explain the experiment and show his original data. The junior scientist didn't deny doctoring the figure. But he did deny intentional fraud. It was an excuse that many at Caltech found hard to believe. "His rationale was essentially, 'I was young and naïve,'" one Caltech source told *Science*.

An urgent meeting of trusted deputies was called at Hood's house. Joan Goverman was there; so was Tim Hunkapiller. Hood had been accused of many things over the years: being overextended, narcissistic, egomaniacal, a bad manager. But no one had questioned his scientific integrity. Science, then and now, was built on trust. Fraud ended careers. Everyone in the lab had something to lose—not just the person accused of committing the fraud. Many had been listed as coauthors on the paper, which raised questions about their own culpability. Not only could a fraud charge end the career of an individual scientist, it could lead to a serious federal investigation, jeopardizing Caltech's ability to win future grant funding. "This was fake data," Goverman said. "This was a big shock wave."

Questions abounded. If this experiment was a fraud, was it an isolated case? Or the tip of an iceberg? Were other people involved, or was Kumar a lone actor? Was everything in the paper wrong? Or only parts of it?

Hood had to act, fast—and with a certain sensitivity. A few years earlier, an NIH scientist, Howard Eisen, had committed suicide while his lab was undergoing a fraud investigation. A fraud inquiry at the MIT lab of David Baltimore, a Nobel Laureate, had become a painful public spectacle. Baltimore, a friend of Hood's, hadn't handled it well. A congressional committee led by John Dingell wanted to know just what American scientists were doing with all their taxpayer support. Baltimore defended his scientists—too quickly, it turned out. He, along with the entire scientific enterprise, was cast in the court of public opinion as suspicious. Baltimore's name was eventually restored. But rumors about what happened in his lab lingered.

Hood knew he couldn't just circle the wagons and fight. Scientific fraud couldn't be downplayed or swept under the rug. But he didn't have much guidance on exactly what to do. "The academic community is still largely working in the dark, without uniform standards," *Science* magazine said.

What should be done about the published paper? Hood wondered aloud: If the conclusion of Kumar's paper had been correct—even if the data underpinning it was fraudulent—was it really necessary to retract the paper?

His advisory team's reaction was swift and unanimous. The paper had fake data; it needed to be retracted. Hood considered their arguments. "It's fair to say he struggled with the question of how much he needed to retract," Goverman said. "Ultimately, he did listen to reason."

Hood alerted the editor of the *Journal of Experimental Medicine* to the fabricated figure. He withdrew another paper that had been submitted to *Cell*, the top-notch journal. Hood apprised Kumar's outside coauthors, and his funders at the NIH, of the situation. He called and retracted the letters of recommendation he had written for the young scientist. Kumar's wife called Hood one night at 1:00 a.m. to say her husband hadn't been able to eat or sleep in four or five days. Hood's wife, Valerie, picked up Kumar the next morning to get some medical help, according to Hood's typewritten notes provided to the Caltech investigative committee. Kumar was put on leave while the investigation played out.

The stress on Kumar and Hood was bad enough. But it didn't end there. The investigation essentially stopped the lab in its tracks. Hood pulled together four committees of postdocs to try to reproduce the experiments published in all four papers Kumar had authored—a year's worth of work. "Hood was on it right away," said Kevles, who wrote a book about the misconduct in Baltimore's lab. "I thought his behavior was exemplary."

Hood informed Caltech faculty of the internal investigation. "It is imperative that the integrity of science and our institution be maintained," Hood wrote in a June 13, 1990, letter. He promised he would "deal expeditiously with this difficult situation." The grim memo went out four years after Hood's triumphant press conference announcing the first automated DNA sequencer.

Caltech's top brass was in full damage-control mode. The new provost, Paul Jennings, wrote a stern memo in all capital letters as a statement to the faculty:

> Among the most important points are that the problem was identified by Lee and his people and that he is leading a vigorous effort to examine the data, reproduce experiments and have ready for the investigation committee any information they might require. It is fortunate in one sense that the scientific result is not a critical one; the scientific impact of the case is not large. However, Lee is a highly visible individual and it seems inevitable that some publicity will soon occur as a result of this

incident. I would appreciate the help of the faculty in minimizing the effects of inaccurate rumors and unwelcome publicity in this sensitive matter.

The biology faculty didn't race to defend the former chairman. If there was ever a moment for Hood's critics to utter "I told you so," this was it. Critics had said his lab had gotten so big that he couldn't possibly oversee his people. It felt like a lonely battle, Hood said.

My perception is I handled virtually everything. I was very much alone in it. I felt the lawyers from Caltech made it clear to me that their obligation, responsibility was to Caltech, and not to me. They'd do anything to help me, but they basically suggested I get my own lawyer. And I did.

Soon, many leading lions of science stood up in defense. Top biologists such as J. Michael Bishop at UC San Francisco and Stanley Cohen at Stanford University sent letters of support. Charles Janeway, an immunologist at Yale, expressed a typical sentiment: "I just want you to know that I have no doubts about your personal integrity, and that I will be happy to help."

The investigation sent a chill through the lab. A once-collegial place was full of suspicion. Postdocs were scrutinizing the lab notebooks of colleagues. Whatever independent research had been on the agenda was put on the back burner. While Hood could hire a lawyer for himself, the younger coauthors of the paper couldn't afford to do the same. Everyone felt vulnerable.

The investigation was long and painful. Certain lab supplies, such as valuable peptides used for the probe, went missing. Rumors about the inquiry—and who might be doing what to interfere with it—swirled around the lab. A year after the initial accusation, the

Caltech investigating committee found Kumar to have committed "research fraud." With two of his four research papers retracted and another withdrawn, Kumar sought a fresh start. He found a postdoctoral fellowship at UCLA, where immunologist Eli Sercarz argued that Caltech acted "precipitously."

The National Institutes of Health conducted its own investigation. Kumar and the NIH's Office of Research Integrity reached a settlement in July 1996, in which Kumar accepted an eighteen-month ban on NIH funding. Kumar maintained that he never intended to deceive anyone. He eventually resumed his career in immunology at a research center in San Diego, and he cofounded a couple of small biotech companies.

But that wasn't the end of suspicion in the Hood lab. The investigating committee, in the course of its work, found that others published questionable results. Like Kumar, James Urban had been working in a hot area of immunology thought to be important for multiple sclerosis. Urban had written a paper, with Hood as coauthor, that contained fake data. Urban "maintained that he had planned to obtain the real data and substitute them for the made-up version before the paper was published," wrote David Goodstein in his 2010 book *On Fact and Fraud*. By the time the Caltech investigating committee uncovered Urban's "serious errors in judgment and scientific misconduct," Urban had moved on to a faculty position at the University of Chicago. He resigned and accepted a three-year ban on NIH funding. Goodstein saw a familiar pattern. Urban and Kumar "were both under intense pressure, and worked long hours, drawing praise for their dedication from Hood," he wrote.

Hood handled the public relations crisis masterfully. He came across as a victim of young people bent on deceit. Reports at the time said little about how his freewheeling, hands-off management style created an environment where fraud could easily occur.

Before the investigation was complete, *Science* magazine picked up the story. Hood tried to get the reporter to back off, but it didn't work. He was nervous. He sent a terse memo to his lab on February 22, 1991, with the subject line "Science Article by Leslie Roberts on the Fraud Cases."

> According to Leslie Roberts, "sources in the Hood
> lab" discussed this case with her. I would like those
> "sources" to identify themselves to me.

No one stepped forward. Despite Hood's objection, *Science* ran the article on March 1, 1991, under the headline "Caltech Deals with Fraud Allegations." The article was measured and careful to avoid pointing fingers or jumping to premature conclusions.

Hood himself was never accused of misconduct. Goodstein was sympathetic. "Hood, who was a careful scientist, tended to believe that everyone else was equally scrupulous, and he therefore was inclined to discount such assertions," he wrote. More generally, he noted that fraud cases could never be completely prevented. "No human activity can stand up to the glare of relentless, absolute honesty," he wrote. "We build little hypocrisies and misrepresentations into what we do to make our lives a little easier, and science, a very human enterprise, is no exception."

But Hood did bear some responsibility for what happened. While the impact on his career was minimal, others in the lab weren't so fortunate. The young postdocs in Hood's lab hadn't made names for themselves yet. They couldn't just get a faculty job anywhere else at a moment's notice. They suffered a lost year checking others' work, when they could have been doing their own science and building their résumés.

The whole process and ultimate outcome was a "disaster," in the words of Ellen Rothenberg, a Caltech biologist and member of the investigating committee. "For one thing, in the end, the guilty come away unpunished," she said. "They get temporarily punished, but they don't stay punished for long. The innocent suffer immensely."

The once-carefree Hood lab suddenly took on all kinds of bureaucratic processes. A new formal review process was installed, which required three people inside the lab to review each paper. The reviewers were expected to roll up their sleeves and look at the raw experimental data, not just skim the summary of the findings. Hood required everyone to keep a bound notebook. After a few critical notebooks in the investigation had mysteriously gone missing, he "made clear it is property of Caltech, not the scientist," according to an article in *Science*.

Formal processes, based on suspicion, drained away some of the joy and some of the adventure that Hood loved about science. Hood was so melancholy that he briefly considered quitting science altogether. He had vision, rhetorical skill, and fund-raising ability. Why not go into politics? He could make a run for the US Senate from Montana.

Everyone around him instantly dismissed the idea as ridiculous. He "would have been a terrible politician," said his son, Eran. The list of reasons was long. Hood had little empathy for other people's concerns, zero patience for committees, and no taste for the art of compromise. All those things should have been obvious from his time as chairman of biology. His earnest and sometimes gullible nature, a gift of his small-town upbringing, would be easily exploited in the shark tank of Washington, DC. And his ego would have gotten in the way.

"I can remember just thinking if he got into a debate, he would jump up at some point and say 'I can prove that you're an idiot.' And he'd go on and list reasons why the other guy was an idiot. That would be his debating style," Eran said. "You could easily bait him to self-destruct."

Hood knew science was where he belonged. So he trudged through his darkest hours. By fall 1991, the fraud case reached its conclusion.

Then Hood got even worse news. Eran, his oldest son, was missing on a mountainside. Climbing with a friend on Mount Scott in Denali National Park in September, Eran got caught in a whiteout snowstorm. They'd been missing for three days. Hood and his wife flew to Alaska. Hood knew his son was a strong climber and skilled outdoorsman, but this was frightening.

Eran and his friend weren't prepared for the conditions. They had only light insulating clothes, one sleeping bag for the two of them, and enough food and water for a day hike. They had to get to low ground, where they could find berries for food, while avoiding grizzly bears doing the same thing. Search and rescue teams, aided by helicopter support, scoured the area. By day six, a ranger informed Hood and his wife that they would continue looking, but the odds of survival were slim. It was "very depressing," Hood said.

While Hood was never the most attentive father, he was immensely proud of his son. He saw the younger man as cast in his own image. Eran was a track athlete, mountain climber, Harvard man. Hood bragged about him to all who would listen. The thought of losing his son was excruciating. Each night, after getting word from park rangers, Hood and Valerie updated their daughter, Marqui, by phone. Hood could do nothing but wait. He was in good climbing shape at fifty-two, but park rules prevented him from joining the search and rescue team.

On day seven, the weather cleared. A fixed-wing aircraft traveling up a river valley spotted the young climbers. Eran's friend had significant frostbite, but they were otherwise healthy. Hood and Valerie hugged their son and cried tears of joy. "We heard them say on the radio, 'We found them! We found them!'" Hood said years later. "It was just spectacular."

In a way, Hood was in need of being found, too. More than anything, he yearned for a place that would embrace him as Caltech had in the early years. He would find it thanks to a budding friendship with a young software mogul from Seattle.

His name was Bill Gates.

Chapter 12

A Fresh Start in Seattle

Seattle in the early 1990s wasn't a world-class city in anyone's book. It bore little resemblance to Los Angeles, the hip and sunny metropolis where Hood had lived most of his adult life. Its culture and history had more in common with Montana, the place where he grew up.

The urban capital of the Pacific Northwest was built at the start of the twentieth century by an Alaskan gold rush. The economy was driven by natural resources—logging, shipping, fishing, and agriculture. Boeing, the airplane manufacturer, became such a regional force that when it laid off sixty thousand workers in the early 1970s, a billboard near the airport famously read, "Will the last person leaving Seattle turn out the lights?" Twenty years later, the city still hadn't entirely recovered from the Boeing bust. Across Lake Washington in suburban Redmond, Microsoft had gone public and was beginning to grow, but it still generated less than $2 billion in annual revenue. The software giant hadn't reached the computing masses; graphical user interface software hadn't yet been popularized by Windows 95. Starbucks barely existed, Costco had yet to challenge Walmart, and Amazon hadn't been born. A few biotech companies had ambitions, but they were mostly anonymous. To the extent Seattle was nationally known for anything in 1991, it was for rain, trees, and grunge music from Nirvana, Soundgarden, and Pearl Jam.

Seattle in those days struggled to support a first-rate university. The University of Washington, with a picturesque campus overlooking Lake Washington, was a big public research university with a respected medical school. But it didn't quite measure up academically to public rivals like the University of California–Berkeley or the University of Michigan. By the mid-1980s, budget pressures had driven away so many faculty and slashed so many programs that UW president William Gerberding complained, "You can't ruin a splendid university in five or six years. ... I don't know how many years it would take, but we're on our way." It wasn't until the end of 1989 that UW started to turn things around with a $250 million community fund-raising campaign.

Partly because of its far northwestern remoteness, Seattle has always attracted entrepreneurial dreamers and supported unconventional ideas. The university was no exception. Bright young faculty could test ideas without much interference. One such bold idea struck the ambitious young immunologist Roger Perlmutter, a Hood protégé.

Perlmutter moved from Caltech to Seattle in the mid-1980s to follow his wife to the University of Washington. She had landed a clinical fellowship in the university's infectious diseases department. With a strong recommendation and some key introductions from Hood, Perlmutter landed an assistant professorship at UW. He did well. Perlmutter, then in his midthirties, persuaded his bosses to let him spearhead a new immunology department. Immunology was beginning to mature as a discipline with its own set of problems and language, no longer a subdiscipline of microbiology. Immunology's profile was rising, and so was Perlmutter's.

That's when Perlmutter heard about Hood's troubles at Caltech. He smelled opportunity.

"Lee was unquestionably the preeminent molecular immunologist in the world," Perlmutter said. Luring Hood would give the department instant credibility. "I was very young to take on the department chair position," he said. "Although my lab had done some good work, I was no Lee Hood. I was looking for marquee leaders."

Perlmutter figured it was a long shot. He knew an offer from Berkeley was in play. But when he called his former boss, he found a receptive ear. Hood, a Montana boy to the core, liked all the hiking, kayaking, and mountain climbing opportunities near Seattle. A bonus: Hood's old college roommate and climbing buddy, Eric Adelberger, was a physicist at UW. Valerie had friends in town, too.

Convinced this was no fool's errand, Perlmutter approached Lee Huntsman, the director of UW's bioengineering department. Huntsman was the perfect ally, a respected scientist running a top-ranked department known for pioneering work in medical ultrasound imaging. Huntsman would see the value of luring a biology and technology dynamo—especially one who would recruit more smart people and amplify everyone's work.

Huntsman also was a UW lifer who understood the polite, non-confrontational, politics-by-consensus way of doing things on campus. A tall, silver-haired diplomat, Huntsman was patient, trustworthy, understated. He spoke softly, and he listened more than he talked. His intellect and maybe even more so his manners gave him influence. Seattleites were ambitious, but they frowned on the attention-seeking and outwardly aggressive displays of power common in big cities like New York and Los Angeles. Perlmutter knew that Huntsman could help persuade key players—both inside UW and in the Seattle business community—to rally around a superstar faculty recruitment. Maybe even the software billionaire in Redmond could be enticed to get on board.

Huntsman warmed to the idea quickly. He scheduled a meeting with Phil Fialkow, the dean of the UW School of Medicine. Fialkow's endorsement was essential. It wasn't just because he controlled a sizable budget and considerable lab space. Fialkow oversaw basic biomedical research and a medical center—one that treated tens of thousands of patients every year. At least in theory, Fialkow could influence professors focused on patient care to work with researchers focused on the basic mechanics of genes and cells. Caltech, while excellent at basic biomedical research, had no medical school. That made it more difficult for scientists to get access to tissue samples, as well as to find like-minded physician/scientists as collaborators. Fialkow knew the value of such collaborations. He made his name as a scientist by studying the genetic basis of leukemia, a cancer of the blood cells—making him a rarity among physicians. Perlmutter and Huntsman knew that if UW could offer Hood the resources and collaborative possibilities of a top-notch medical school, they might have a shot.

Captivated, Fialkow thought even bigger. "Let's create a whole new department [around Hood] and put UW School of Medicine resources behind it," he told Huntsman. So in the fall of 1990, the recruiting began. To reel in a big fish like Hood, they had some attractive bait. But they also needed money.

Mary Gates was a highly respected member of Seattle society. She cared deeply about the University of Washington, serving on its board of regents for eighteen years. William Gates Sr., the other half of the power couple, was a partner at a prestigious law firm.

For years, UW administrators had tried to talk the Gates's son into donating to his hometown university. Bill Gates never attended UW. But as teens, he and his friend Paul Allen got access to huge university computers that helped spark their imaginations.

About fifteen years after the founding of Microsoft, Bill Gates was still concentrating on his company. Yet pressure was mounting to start giving away money. At thirty-five years old, Gates was already the second-richest man in America, with a net worth estimated at $4.8 billion by *Forbes*. (By 2015, his fortune had ballooned to $79 billion.) "There was a hope that Bill would get interested in the UW at some point," Huntsman said. "But nothing had really resonated yet."

Huntsman had a feeling this could be it. He brought the plan to the man himself. Sitting in a luxury box—watching a football game between the UCLA Bruins and the hometown Washington Huskies on November 10, 1990—Huntsman told Gates about Hood. He spoke of Hood's interest in biology and technology. Gates had already shown some interest in biotech. A few months earlier, Gates had invested $5 million in Icos, a Seattle-based drug developer led by CEO George Rathmann, a star executive from biotech giant Amgen. Icos raised a then-record $33 million initial round of venture capital.

Gates listened. He asked for some of Hood's scientific papers. Excited that they had Gates's attention, Perlmutter invited Hood to give three lectures at UW in April 1991, as part of the distinguished Jessie and John Danz lecture series. Gates attended all three talks, sitting near the front of the Kane Hall auditorium. Hood couldn't miss him.

The Danz lecture series was essential to this academic courtship. It attracted distinguished scholars, in its official description, "who have concerned themselves with the impact of science and philosophy on man's perception of a rational universe." The connection between science and society was right in Hood's wheelhouse.

Hood's initial lecture bore the expansive title "Biotechnology, the Genome Initiative, and Medicine of the Twenty-First Century." Perlmutter wanted to make sure Hood's first impression was memorable. In his introduction, Perlmutter noted that Hood was a native of Montana, "a local boy," and "a scientific visionary of extraordinary breadth and scope." For nearly five minutes, Perlmutter poured on the praise. He portrayed Hood as the leader of a scientific revolution.

> During the past couple of decades, Dr. Hood has championed a research program that has had an enormous impact on the way in which biomedical research is conducted in the United States and elsewhere in the world. In particular, he assembled the tools necessary to apply advanced thinking in electronics and instrumentation to the analysis of biologically important molecules. By developing improved technologies and an ever-more sophisticated approach, he was able to define the elegant strategies that are employed by all vertebrate organisms to generate specific immune responses. ...

> Beginning in 1983, he began developing a novel, high-technology approach to what some call the greatest adventure of self-discovery yet contemplated by mankind. That is, the complete structural elucidation of the DNA sequence that defines our species. ...

> Dr. Hood and his colleagues have formulated strategies that promise to revolutionize medical diagnosis. Beyond that, he has argued convincingly that the information gained through these studies will illuminate mechanisms of molecular evolution and the biochemistry of all living cells.

Bathed in applause, Hood stepped to the podium. As was his habit at the beginning of a lecture, he smiled, made eye contact with several people in the audience, and cracked a joke to show he wasn't completely full of himself. "Thank you very much, Roger, I think you have just given my lecture." The audience was at ease.

Hood, settling in with his slide show, gave one of his most brilliant and persuasive talks. His lecture was understandable for anyone at about a tenth-grade level of comprehension. He could see the big picture and explain it in plain English, in ways that almost no other biologist could. Hood made a case for the "genome initiative" and the payoff to come for human health.

The breadth of topics he covered was remarkable. He started with basic information about what genes are and what they do, how certain regions on the genome control which genes are expressed in certain tissues at certain times, and how the genetic information gets transcribed into the proteins that are the workhorses of the cell.

He talked about the mystifying "protein-folding problem." Just because a scientist knew the gene sequence, and the sequence for its corresponding protein, didn't mean he knew what the protein really did or how it functioned. The protein's function depended on how it folded into a particular 3-D shape. That unpredictability made protein folding an entirely separate and more difficult problem from sequencing the genome. It would take biologists decades to chip away at the relationship between a protein's structure and its function.

Hood raised all kinds of interesting questions for the audience to consider. Does anyone have the right to patent a gene? (That decision would have profound implications for the fledgling bio-technology industry.) Should employers or insurers have access to an individual's genetic data? How would medical schools teach doc-tors to practice medicine based on genetic information and main-taining wellness? How could K–12 science education prepare a new generation of scientists for this work, as well as educate citizens to think about its ethical dilemmas? How much should these precise genetic diagnostics and therapies cost? (Imagine a drug that could prevent an autoimmune disease, such as rheumatoid arthritis; even at a huge expense, it could raise the productivity of millions of citizens and thus be cost-effective.) In many cases, the societal issues that Hood raised in such clear, vivid terms were ones that scientists and policy makers continued to wrestle with decades later.

As in all of his best talks, Hood never condescended. He treated listeners with respect, as if they were fellow travelers on this great intellectual journey. Instead of using stilted academic phrases like "one must assume" or "it does appear correlated with," he used conversational language. Sometimes, he flattered his audience by saying he assumed they knew basic things about biology. (He'd explain it anyway.) When talking about things he and his lab were doing, he frequently used the inclusive pronouns "we" and "us" to suggest the audience was there with him.

He outlined the big problems of biology and his vision for how technology could solve them. He used words like "exciting" and "wonderful" and "powerful" in the repetitive patterns used by all great orators. He used the word "fundamental" forty times in a seventy-five-minute speech, always in the context of how the genome initiative would advance some fundamental understanding of life. The journey of discovery was grandiose. Hood was brimming with his Boy Scout enthusiasm:

> What's wonderful about this informational
> macromolecule, DNA, is the information transfer
> that 4.5 billion years of evolution has generated
> to take the one-dimensional information from
> our chromosomes, which are basically a hundred
> thousand units of genes or so, and translate it

through a complicated process into the three-dimensional information of our bodies.

The final outputs of that informational process are proteins. These are linear chains that are made up of twenty different amino acid subunits. The order of these subunits specifies how a particular protein, made by a particular gene, pulls together in three dimensions to form a molecular machine. These molecular machines give our body size and shape and form, and they catalyze the chemistry of life.

One of the challenging puzzles that remains before us is: What are the rules of how we translate one-dimensional information of DNA and genes into the three-dimensional information of proteins? As you'll see, the genome initiative will provide us with some very powerful tools for attacking this fundamental question.

If the Human Genome Project were to go forward, Hood said, it would provide a path to answer some of the big questions in biology—and raise even more questions. These included, in his words:

• What are the means by which these hundred thousand genes specify their actions at precise points of the developmental stage in particular kinds of cells?

• What is the regulatory machinery that turns on each of these genes at the right time, in the right place, and in the right quantity?

• What is the nature of the molecular choreography that permits the thousands of genes expressed in a muscle cell to be differentiated from the thousands of genes expressed in a brain cell or a skin cell?

In Hood's view, the purpose of the genome project wasn't necessarily to answer all those questions, at least in the beginning. It was worth doing so that scientists could begin to think ambitiously. "The genome initiative will provide very powerful tools so we can begin to attack," he said.

Just getting the sequence for starters was a "gargantuan task." Biologists would be busy with the work of interpretation and deciphering the information for at least a century, he said. He wasn't promising a quick fix.

Obviously, Hood said, sequencing the genome would require big advances in computer processing power and software. That's where Bill Gates entered the picture. Hood offered up a tantalizing vision for how he, the biologist, could connect with Gates and other leaders on the technology side. He said:

> What biology can do at a given point in time is pose
> challenges—frontiers beyond which it can't look.
> It is then the challenge of technology to develop
> techniques that can pass over those barriers, over
> those challenges, and in doing so, the technology
> will then open up new vistas for biology to explore. I
> see these two things as being synergistically applied,
> back and forth, one to another ... clearly, what we
> need to be able to do in the future is to really bring
> people from hard-core computer science, applied
> math, physics, chemistry, and engineering into this
> whole process. That's an incredibly challenging kind
> of problem. How can you actually, really, go about
> doing that?

Hood had already started going about doing that. At Caltech, he ran a program in which graduate students committed to becoming experts in biology while also studying an adjacent field, such as computer science. The students would have a mentor in each field and become "hybrids." Ideally, the students could serve as translators between their mentors, who didn't usually talk to each other or understand each other's fields. If you could bridge those gaps, and really work at the nexus of these different disciplines, Hood thought, then you could make breakthroughs.

As the first Danz lecture wound down, Hood aimed his interdisciplinary message at one person in the audience: Bill Gates.

"I would argue one of the major opportunities for industry in the future is going to be at the intersection of computer science and biology," Hood said. "In fact," Hood said with a lilt in his voice and an upward jab of his index finger, "if I had to pick any discipline, apart from chemistry, that's going to make really fundamental contributions to biology over the next fifteen to twenty years, it would be computer science."

Gates was intrigued. "When I first saw Lee at a lecture here at the UW … he was absolutely right," Gates recalled. "He's a very brilliant guy who is kind of a Pied Piper."

Over the course of his two-week visit to Seattle, Hood gave three Danz lectures, plus smaller seminars almost every day. He visited the physics, computer science, and medical genetics departments. At every stop, he could drill deep into esoteric scientific questions, and just as easily zoom back out to describe the big picture, framing the big questions in clear language.

The world, Hood cautioned, might have a hard time keeping up with all the changes. "But it will be exciting," he said in closing the first Danz lecture. "I'd argue we'll learn more about man, and what he is, in the next twenty-five years than we have in the last twenty-five hundred."

Soon, the campus was buzzing about the visiting biologist from Caltech. "When you first meet Lee, you're overwhelmed with his intellect," recalled Ed Lazowska, a UW computer science professor. "It's not that he's overbearing, it's just that it's evident. He looks at you with those eyes, and you understand how much is behind them." In his field, Lazowska remarked, there were problems that were trivial and problems that were profound. Hood, he said, was a rare outsider "who knew the difference" and could see linkages between his discipline and others.

After the third and final Danz lecture, it was time to talk business. Gates and Hood, joined by Huntsman and Fialkow, went out for dinner at the Columbia Tower Club, a fancy restaurant on the seventy-fifth floor of the tallest building in Seattle with views of Puget Sound and the Olympic Mountains.

Hood, fifty-two, and Gates, thirty-five, hit it off. The dinner stretched a full four hours. Gates and Hood found they could speak each other's language. Biology was based on an underlying language of code, much like computers. Instead of zeros and ones, the biological code that governed much of life was written in billions of complementary As and Ts, Cs and Gs. "It was a fascinating experience," Hood recalled. "Bill asked me an infinite number of questions about science." His answers? "It was just what you'd expect from Lee," Huntsman said. "The first thing he talked about was the vision. The second thing he talked about was the vision. And the third thing he talked about was the vision."

Their talk was so absorbing, so intense, that Gates and Hood appeared to forget their dining companions altogether. It was no surprise. Walter Isaacson once wrote that Gates had "an awesome and at times frightening blend of brilliance, drive, competitiveness, and personal intensity." The same could have been said about Hood.

Huntsman and Fialkow sat quietly through dinner, observing these two. The UW leaders joked afterward that about twenty minutes into the dinner, Gates became convinced that his job was to recruit Hood, and Hood became determined to win the backing of Gates. Hood said later:

> I think my mind was made up very quickly after that dinner. Bill Gates was an interesting, intellectually challenging person. I'd never met anybody quite like him. There wasn't anyone on the board at Caltech that was the same dynamic, driving, inquisitive, and intense kind of person. It was terrific. He just got things, and could get to the heart of the matter so quickly. We didn't agree on everything, but I think we respected where we have differences.

A few weeks after Hood's visit, the immunology faculty met to talk about any concerns. For more than an hour, they discussed the impact of "big science" on the university, and whether Hood could juggle all of his administrative duties while overseeing graduate students. Then there was the unsettling case of the fake data.

To UW, these were minor issues—a moment of doubt in Fialkow's mind, but nothing more. Hood's intellectual impact on the department "would be in every respect beneficial," Perlmutter wrote to Fialkow. Hood would help recruit first-rate faculty and graduate students to Immunology and the broader School of Medicine. "I wanted you to know that the Department of Immunology enthusiastically supports this recruitment effort, and that we are keen to have Lee as a colleague," Perlmutter wrote.

When Hood expressed some unhappiness with the space he was being offered, Perlmutter deftly proposed a solution: he'd grant Hood a joint faculty appointment in his department, Immunology, and throw in a small amount of his lab space.

With the best possible offer in place, Fialkow flew to Los Angeles in fall 1991 to woo Hood one last time. Fialkow assured Hood that he would get the space he wanted. And that Fialkow would personally protect Hood's back as needed. That clinched it. Still wounded over the mutiny at Caltech, Hood wanted to feel some love. He drew up a formal request for $25 million to create a new UW Department of Molecular Biotechnology.

It was an audacious request for an audacious goal. The department would be the first of its kind, eschewing the silos favored by scientists and drawing from several disciplines: biology, mathematics, physics, and computer science. It was modeled after what Hood had built at Caltech, but this time it would be formally sanctioned and approved in the academic hierarchy. In a handwritten note to Fialkow shortly after the dinner with Gates, Hood explained:

> Bill struck me as an individual who thinks big and likes to strike out in innovative new directions with appropriate resources. I would like to consider the possibility you ask for $25 million initially, which strikes me as very much in keeping with the bold new venture we have in mind. In my dealing with wealthy and smart Caltech trustees (e.g., Beckman), he is more interested in doing new projects correctly than in worrying about the bottom line of the initial cost. ... I hope you take my comments as an indication of how excited I am about this opportunity.

Gates found the full vision striking indeed. He agreed to provide half of what Hood asked for: $12 million. It was more than enough to start. All parties were thrilled. "It was a chance for Hood to take his game to a larger arena, with larger impact at a medical school," Huntsman said. "We were quite sincere about this being a good place for Hood and his team. Simultaneously, we thought it would bring a new zone of excellence into our midst. It was going to be good for everyone."

A deal was done. Hood would become the William H. Gates III Professor of Biomedical Sciences and the founding chair of a new Department of Molecular Biotechnology at the University of Washington.

The agreement called for Gates's money to be divvied up several ways. About $3 million went into an endowment; investment income would pay Hood's salary, initially set at $150,000 a year. An additional $3.5 million went for laboratories, $2 million for equipment, $1 million for office support, and $2 million in seed money for biotech research projects that could be doled out largely at Hood's discretion. With that base of support, Hood could recruit about ten faculty members.

Faculty members would have to be somewhat daring to join a multidisciplinary research center like this. For all scientists' talk of tearing down walls between areas of expertise in academia, few wanted to actually do it.

Scientists knew that having an outside perspective could result in a great innovative leap. Linus Pauling, for example, brought formidable expertise in chemistry to bear on biological problems, which helped him see things about sickle cell anemia that others hadn't. Still, moving across disciplines felt too scary. It meant moving outside your own comfort zone, your own area of expertise. What Hood had in mind wasn't "real biology," traditionalists sniffed. But there were people who thought molecular biology sorely needed an injection of new ideas.

Maynard Olson was one such scientist. Olson, at Washington University in St. Louis, emerged in the 1980s as a player in the new field of genetic sequencing. A Caltech alum trained in chemistry, he questioned much of the dogma in biology. He noted:

The scale with which the work is done in a molecular biology lab—the way things are pipetted from one place to another and then centrifuged, for example—has changed very little over the past 20 years. But during this time there has been a spectacular progress in many areas of technology. You don't find this parochial view in physics or chemistry, where people understand that if you want to learn more about molecules or nuclei or whatever, you need better methods.

Hood was running toward a great opportunity: Gates's money, the chance to build a radical new department based on better tools, new methods, a hand-picked team, the blessing of the medical school, widespread community enthusiasm. Joining UW gave him the freedom to operate that many scientists can only dream of. At the same time, he was running away from an unhappy situation. "I would like to move as soon as possible after the decision is made," Hood told UW's medical school dean.

Hood knew this news would be a bombshell, and he kept a tight lid on it. Back in Pasadena, Hood had told only his wife about the new job. He asked UW officials to send their written correspondence to his home, not his lab, where prying eyes might wonder about all that mail from Seattle. It had been ten months since he turned down the Berkeley offer, and the fraud investigation had just wrapped up. Things were starting to get back to normal when Hood broke the news. Before the formal announcement, Hood called Caltech President Tom Everhart. Despite his grievances with Caltech administrators, Hood didn't want his bosses to learn of his departure on the evening news. "I felt I owed them that," Hood said.

Hood left Caltech with bittersweet feelings. He had come of age there as an undergraduate and graduate student. As a young man, he proudly wore his Caltech letterman's jacket when he went home to Montana. He spent twenty-two prime years on its faculty. It was the crucible that forged him into a scientific leader.

In a letter to Caltech benefactor Arnold Beckman, Hood wrote: "I have the sad duty to tell you that I have decided to leave Caltech. This is a decision about which I have agonized on and off for the past several years. I am delighted to report that I have an opportunity at the University of Washington Medical School that truly excites me." Hood then told Beckman about some of the administrative problems he perceived at Caltech but said he hoped to maintain scientific ties.

News of UW's hire was trumpeted by the local, national, and science media. The university held a press conference on Monday, October 7, 1991. UW President William Gerberding made welcoming remarks. He was joined by Bill Gates in a jacket and tie, sporting a mop of hair and big, nerdy glasses. Hood, also wearing a suit and tie, was his usual clean-cut self, looking more like a corporate vice president than a disheveled scientist. It was three days before Hood's fifty-third birthday.

"William Gates's Gift Spurs Shift by Star Biologist," the *Wall Street Journal* headline said. The story began: "Microsoft Corp.'s billionaire chief executive officer, William Gates III, bankrolled a major coup by the University of Washington." The nation's leading business newspaper quoted Gates as saying, "The whole area of [molecular biology] fascinates me. After computers, it's the most interesting, most impactful area in research today." Gates and Hood both said new "computational tools" would be needed to make sense of the information flowing from the Human Genome Project. UW's gain was Caltech's loss. "Lee Hood has been a visionary leader in biology at Caltech, and we shall miss his many contributions," Everhart said in a statement.

In Seattle, UW President Gerberding was over the moon. "Today, October 7, 1991, is a momentous day in the history of this university," he said. Not only was Hood bound to bring the University of Washington up to the major leagues of biotechnology, but this was Gates's first large act of philanthropy. It was, at the time, the biggest individual gift ever to the university. Hood's optimism jumped off the page in the *Seattle Times*. "We hope to change the world of medicine," Hood said. Gates called Hood a visionary. A week later, UW's president was still beaming. He wrote a thank-you letter to the young software wizard. "The glow over this institution because of your gift lingers on," Gerberding wrote. None of the news articles discussed the mutiny that cost Hood his chairmanship at Caltech; only one touched briefly on the recent fraud.

Inside the Hood lab, "there was complete pandemonium" for weeks, said John Yates, then a Caltech postdoc. Hood called an all-hands-on-deck meeting to announce the move to Seattle. He wanted to bring some, but not all, of his troops north. By now, weary from the fraud investigation, the lab had shrunk by half, to about fifty people. A series of individual meetings followed, as lab members talked through their options.

Moving to the University of Washington struck many as odd. "Some people were enthusiastic, and some were really unenthusiastic," Hood said. Those with an interest in technology, computer software, and the nascent field of genomics knew they were more likely to fit into the new department. Others, particularly in immunology, saw the writing on the wall: Hood's interest was waning.

Some of those who wanted to stay with Hood didn't want to leave family, friends, or the California sun. "Lee had to do a lot of persuading," said Lee Rowen, who eventually accepted the boss's offer. "I was not particularly drawn to Seattle. I had been there once. I didn't see what the hype was about this place. I figured I'll get seasonal affective disorder. It's dark, it's rainy, it's cold."

More than that, though, the scientists wondered: What will happen to my career if I leave a world-class institution for a mediocre state school? Like any big state-supported university, the University of Washington had a public mandate to make education available to a large number of state residents. It couldn't be as selective as a private institution like Caltech. Hood didn't deny there was a difference, but as usual, he spun it into a positive. Hood "readily acknowledges that the University of Washington is not among the top five research universities in the U.S.," wrote the *Scientist* magazine in March 1992. That was part of what attracted him, he said. "It would be very hard to persuade the Stanfords, the Caltechs, the MITs to build the kind of department I have in mind," Hood said. "They are already very well-off, and they aren't going to make major perturbations to bring in new opportunities. It is the universities that really have a major commitment to improving themselves that are willing to take these chances."

Still, this wasn't the backwoods. UW did have some world-class people and departments. Washington's longtime Democratic US senator, Warren Magnuson, was a power player who became close to John F. Kennedy and Lyndon Johnson. Biomedical research was one of Magnuson's national causes, and he wanted Seattle to be one of the nation's biomedical-research hubs. "Maggie" helped create the highly regarded Fred Hutchinson Cancer Research Center in Seattle in the 1970s. The huge UW Medical Center, the only medical school serving a five-state region, still bears his name.

By late 1991 when Hood began attempting to persuade his postdocs to join him, these efforts were starting to pay off. UW Professor Hans Neurath was a prominent biochemist. One of his best students was Bill Dreyer, Hood's graduate adviser who made a signature accomplishment with the early protein sequencers. The UW biochemistry department under Neurath produced three future Nobel Laureates: Edwin Krebs and Edmond Fischer, who stayed at UW, and Martin Rodbell, who left for the National Institutes of Health. Arno Motulsky, a leader of medical genetics, was a faculty stalwart. So was

Leland Hartwell, a yeast geneticist and future Nobel Laureate. UW was part of the first wave of recombinant DNA experimentation in the 1970s. During that time, Benjamin Hall developed novel means for expressing protein molecules in yeast cells; it became a biotech industry mainstay.

Seattle scientists didn't worship pure theory; they also embraced practical applications of medical technology. Early forms of kidney dialysis, ultrasound imaging for prenatal care, and lifesaving bone-marrow transplant techniques were all pioneered in Seattle. The Fred Hutchinson Cancer Research Center had become a magnet for bright and entrepreneurial faculty. A couple entrepreneurial immunologists from "the Hutch" started a company, Immunex, that went on to develop one of the best-selling drugs of all time: Enbrel for rheumatoid arthritis and other autoimmune diseases.

Seattle presented its advantages for a biological frontiersman. Unlike the East Coast, with its hierarchy, Seattle was a place where pioneers could take big risks without bowing to tradition. It was a welcoming environment for freewheeling, independent spirits.

These were the arguments Hood made, over dinner at his house, to his postdocs and graduate students. That plus upward career mobility: coveted faculty positions for a few. The recruits included future UW genomics leader Debbie Nickerson, the immunologist Joan Goverman, computational biologist Tim Hunkapiller, technologist Alan Blanchard, biochemist Kai Wang, proteomics researcher John Yates, DNA sequencer Robert Kaiser, and genomics researcher Lee Rowen. Hood's entourage from Caltech would number twenty-four—nearly half of the lab.

Hood also needed some big-name researchers from other schools to give his new department cachet. He didn't bother recruiting any fellow faculty from Caltech; most of those bridges were burned. He wanted a key player in the Human Genome Project: Maynard Olson.

Olson had served with Hood on an influential National Science Foundation committee, known as the Alberts Committee, which paved the way for the Human Genome Project in the late 1980s. Recruiting Olson away from Washington University in St. Louis would hopefully cause other dominoes to fall in UW's direction. Hood pitched his vision of building a new department from the ground up, with an emphasis on enabling technologies and on interdisciplinary collaboration.

"Irresistible," Olson called the opportunity. He walked away from a comfortable amount of support from the Howard Hughes Medical Institute to take the new job. *Science* magazine called the Olson recruitment a coup. Soon after, others arrived. A power couple from Lawrence Livermore National Laboratory with science and technology skills—Barbara Trask and Ger van den Engh—signed on. Another star, Ruedi Aebersold of the University of British Columbia, followed. Phil Green, a colleague of Olson's at Washington University and a leader in computational biology, came later. One of the Caltech recruits, Debbie Nickerson, said so many bright minds gathered around Hood because he could make things happen. Hood, she said, had "natural leadership and tremendous enthusiasm. It infects us all."

"We got everyone we wanted," Hood boasted to *Science*.

With his trademark intense gaze, Hood posed for the cover of the *Seattle Times* Sunday magazine. His eyes revealed a sort of laser-sharp aggression that sometimes startled people. He didn't fit the stereotype of a mild-mannered biologist in a white lab coat. He looked more like a linebacker, or a Navy SEAL. "Designer Genes: Leroy Hood Is the Country's Leading Geneticist and the UW's Got Him," the headline gushed.

Life was good. The change of scenery, from palm trees to evergreens, was invigorating. Hood was brimming with ideas for his new department. He met all kinds of interesting new people—including techies from Microsoft—who asked how they could help. Mary Gates, the powerful UW Regent, took a liking to Hood and his wife, Valerie. She introduced them to many of her wealthy and influential friends in Seattle society.

Hood was becoming wealthy himself. He collected royalties on each instrument sold by Applied Biosystems, and those royalties began adding up to millions of dollars for Caltech and the individual inventors. Amgen had grown into the first big biotech success story; Hood still held all the shares he'd received as a founding scientific adviser.

And, by sheer luck, Hood got a windfall from a company he was barely involved in. Along with Irv Weissman of Stanford and David Baltimore of MIT, Hood was a scientific adviser to a Bay Area biotech start-up called Systemix. Hood's relationship with the company was marginal at best, but executives used his name to attract investors and partners. That helped maneuver Systemix into position for a megabucks sale. Just a couple months after Hood got the UW job, Switzerland-based Sandoz (now part of Novartis), paid $392 million for 60 percent of Systemix. The company never made a profit and never won FDA approval to sell any of its products. But biotech stocks were hot that year, and the company sold at the right time.

The adviser with the biggest stake in the company, Irv Weissman, made the most—a reported $20 million. To celebrate, the three scientists spent their stock-market winnings on a Montana ranch near the Bitterroot River in southwestern Montana. It was a stunningly beautiful setting, surrounded by mountains, a couple hours away from where Hood was born in Missoula. It was an ideal spot for scientific retreats and family vacations.

At that secluded spot, Weissman, Baltimore, and Hood would talk about science around the fireplace, on hiking trails, and along trout-filled streams. They talked about Hood's rare opportunity to build something exciting and new at the UW.

Hood was rich, but he was still cheap in some respects. He still wore a plastic wristwatch. He still dressed in khakis. He drove a beat-up Toyota Camry for twenty years, before trading it in for an environmentally friendly Prius. He had little interest in fine food or wine. Weissman and Baltimore became wine aficionados, building a well-stocked cellar at the ranch. They slapped red stickers on certain bottles: off-limits to Lee. They didn't want the fine wine to be wasted on someone who didn't appreciate it.

Hood relaxed into this setting reminiscent of his youth. He would sometimes do the same humble camp chores he did as a boy helping out his granddad Roy. Carl Weissman, Irv's son, remembered Hood donning an apron to wash dishes. "In my lifetime, I've never seen my father, or David Baltimore, wash a dish. But I have, on many occasions, seen Hood alone in the kitchen after all three families have eaten, doing all the dishes, on his own," Carl Weissman said.

Hood preferred spending his money on that essential extension and envelopment of a person: his home. He and Valerie, as empty nesters, went to work building their dream house. The home along Lake Washington, a short jog from the university, cost about $4 million. Hood hired a prominent local architect and requested the finest handcrafted wood and materials. The result was classic Northwest: soaring cathedral ceilings, fireplace, a sweeping deck with lake views, a spiral staircase, and tall east-facing windows that soaked up the sunrise.

The home was a showcase. With 6,500 square feet, plenty of space was devoted to a growing art collection. The centerpiece was an 1,800-pound mahogany sculpture with swooping curves that sat in a squat formation. Hood bragged that it took six men and a barge to move.

"We wanted to create a house that was a piece of art. We largely did that," Hood said. "The house was just spectacular. You could sit there in the morning, watching the sunrise come up over Lake

Washington. You could look up at the ceiling and see these wonderfully arching wood beams."

Hood and Valerie often played the role of hosts. Visiting scientists stayed at a cottage on the property. Hood invited members of his department for dinner. Occasionally, they hosted political fundraisers, usually for centrist Democrats. They loved showing off the house.

Valerie saw the relocation as a chance to redefine herself. By the time of the Seattle move, she had grown weary of all the focus on her brilliant and indefatigable husband. She legally switched back to her maiden name. In Seattle, she would be known as Valerie Logan or sometimes Valerie Logan Hood, but not Valerie Hood anymore. She remained Hood's loyal life companion, and served as his most trusted adviser and confidant, often by his side at important meetings or traveling with him. But with Hood constantly on the road, she had her own life, her own circle of friends. As a petite, quiet woman with a bob of reddish brown hair, she was often overlooked and underestimated. But she was also fiercely independent and one of the few people who could tell Hood when he was wrong. She wanted to build on her own independent identity.

Hood was happy to see Valerie reinvigorated by the change of scenery. He was thrilled to be embraced by a smaller community that was buzzing about his arrival. Businesspeople wanted to meet the man who had the Midas touch with Amgen and Applied Biosystems. Educators were open to his and Valerie's ideas for revamping science education. Politicians wanted to know how to diversify the local economy through biotech.

And he was becoming friends with Bill Gates.

As head of the UW Department of Molecular Biotechnology, Hood regularly briefed the department's founding benefactor. Sometimes, Gates would invite colleagues like Microsoft chief technology officer Nathan Myhrvold to join the freewheeling biology

Q&A sessions, which were usually held at Microsoft. Hood hit it off with high-tech types. They were curious about his field, occasionally seeking inspiration for their work in artificial intelligence. Hood enjoyed telling them about the human immune system and its astonishing adaptive intelligence capabilities. In turn, Hood thought computer people could help make sense of the DNA code, which looks to the human eye like gibberish.

"The thing that fascinated Bill [Gates] and myself and others in computers is that the genetic code is kind of a programming language that kind of executes a program," Myhrvold said. "I say 'kind of' because it's not exactly what happens, but pretty much. It's interesting to see the degree to which there are parallels.

"Some biologists you talk to about this are so into all the minutiae and the details that a nonspecialist can't have a conversation with them. It's certainly not true for Lee Hood. You could actually talk to him," Myhrvold said.

Gates and his girlfriend at the time, venture capitalist Ann Winblad, once traveled to Brazil, lugging around James Watson's classic textbook *Molecular Biology of the Gene*. They struggled, and Winblad bought a glossary to help them decode it. You never needed a glossary around Hood. "Lee is such a good teacher," Winblad said. "He really loves explaining things and telling people things. Once you intersect with him, you can ask him questions anytime. He's very open to smart people of every kind, especially people who are interested in learning."

Gates was interested in learning. "I was studying a lot of biology at the time, so Lee was educating me," he said.

In August 1993, Gates and fiancée Melinda French invited Hood and Valerie to join a dozen friends on a three-week safari to central and eastern Africa. It was Gates's first trip to Africa, the continent that would later consume much of his attention as a philanthropist. Among several talented people on the trip was Donald Johanson, the paleoanthropologist famous for discovering fossil remains of the primate "Lucy."

Gates and Hood bonded around the campfire on the Serengeti plain. Late one night, wild animals baying under the moonlit sky, Gates and his new biologist friend discussed the lack of computers and Internet connectivity in poor countries, which some later called the "digital divide." Hood talked about problems in biotech that he hoped computers could solve, such as helping to select the best candidate molecules for drug development. Pharmaceutical companies—then and now—waste huge amounts of time and money on experimental drugs that look promising in a petri dish but later fail in trials with animals or people. Hood said he had cofounded a start-up called Darwin Molecular to work on this problem. (Gates invested $5 million, as did Microsoft cofounder Paul Allen. When the company was sold for $120 million a few years later, Gates and Allen tripled their investments. Hood pocketed $2 million, although his boss, the medical school dean Phil Fialkow, ordered him to give all the proceeds to charity to avoid any appearance of conflict of interest with his research. "I was pretty annoyed," Hood said.)

One night after dinner, the eclectic group held a talent show. Hood played a song on his recorder, Winblad recalled. And someone started a game of bridge, which Hood didn't know how to play.

"It's really hard to play bridge. But Lee goes into his tent for a couple nights, reads the book on how to play bridge, and is suddenly a bridge player," she said.

"He really has a very unique mind," said Winblad, who struck up a lasting friendship with Hood on the safari. "Many scientists are extremely focused, and they stay in that focus; they become almost like savants at what they do. Lee is practically the opposite. He can become instantly savant at anything."

"He's a guy who doesn't sleep much," Gates said.

Hood was too excited to sleep. He had new frontiers to explore.

The new UW Department of Molecular Biotechnology was taking shape. He could get laser focused on science and technology again. The future looked bright. "There was an enormous sense of exhilaration and a feeling that we were going to do new and important things and change the world," Hood said.

He was his old whirlwind self again, ready to go.

Chapter 13

New Frontiers, Bold Promises

Racing along Lake Washington during his solitary morning runs, Hood imagined a new agenda for the 1990s. Biology was moving fast. The Human Genome Project was transforming biology from solo act to team sport. DNA sequencers still weren't Ferraris, but the data was coming faster than ever. No one person could collect all the DNA data and figure out what it meant for health and disease. No single scientific discipline could, either.

Hood's bold new Department of Molecular Biotechnology took people who spoke different languages—biology, chemistry, mathematics, computer science, physics. Putting them in one building was easy. Getting them to work together was hard. An entrepreneurial attitude, a willingness to break out of one's comfort zone and try new things, was just as important as the person's expertise. It was going to take entrepreneurial spirit, too, to dream up and design the next technologies in the genomics era. Applied Biosystems, in the San Francisco Bay Area, and other research groups around the world focused on continually improving the DNA sequencers. Hood's team directed some energy toward advancing DNA sequencing technology, but Hood also sought out experts who could develop other types of enabling tools for the modern biology lab.

• Alan Blanchard, recruited from Caltech, repurposed ink-jet printers so that they could print out 3-D oligonucleotide molecules quickly and under computer control. That helped scientists analyze the way genes are expressed—turned on or off—in a tissue sample.

• Phil Green, "the absolute top in the world" at applying math to molecular genetics, according to MIT star Eric Lander, came from Washington University in St. Louis. Green crafted software that could assemble a genome by matching up overlapping ends of the short DNA stretches that were fed through the sequencers. Without the software, it would have been like trying to assemble a jigsaw puzzle with a billion pieces scattered on the floor.

• Ger van den Engh, from Lawrence Livermore National Laboratory, developed new technology for high-speed cell sorting. The instruments, known as flow cytometers, helped scientists look into a blood sample and count the number of cells carrying a distinct marker. Monitoring the progression of HIV infection by sorting and counting the cells bearing CD4 and CD8 markers was just one high-profile use of flow cytometry.

• Barbara Trask, also from Lawrence Livermore, worked on advanced developments in fluorescence in situ hybridization (FISH) technology. It was a way of using fluorescent probes that bind only to specific DNA sequences located on chromosomes. This technology was fundamental to determining the presence or absence of a gene, or the presence of malformations, deletions, and rearrangements. The information from FISH technology was widely used in medicine and by genetic counselors working with patients who had rare diseases.

• Debbie Nickerson, from Caltech, tracked subtle variations in genetic code known as single nucleotide polymorphisms, or SNPs (pronounced "snips"). Nickerson studied how those subtle variations in a single building block of DNA could manifest themselves in observable differences in people. Connecting the dots between SNPs and differences that a physician might observe (known as phenotype) would keep scientists busy for decades.

• Ruedi Aebersold, from the University of British Columbia, developed a tagging system to better measure the relative quantities of proteins in one sample compared with another. This improvement to mass spectrometry was essential to proteomics—a field devoted to large-scale analysis of the distinct proteins that arise from the genetic code.

• John Yates, another Caltech import, also played an important role in mass spectrometry and proteomics, which many saw as the next step for a post-genome world in the twenty-first century.

• Olson, the big genomics fish from Washington University in St. Louis, continued his work on gene sequencing, which eventually led to his running one of the sixteen international genome sequencing centers that were part of the Human Genome Project.

"Nobody could accuse Leroy Hood of thinking small," *Science* magazine said. The *Los Angeles Times* called Hood "a key pioneer in the human-genome revolution."

Hood predicted that by 2016, Americans would carry their personal genomes on a digital card. "I can't imagine that won't be true," Hood told the *LA Times* in a 1996 feature story. "I think that's absolutely a given. Your entire genome and medical history will be on a credit card. You just put it in [a computer] and a physician will instantly know what he's dealing with. Physicians will really then have to look at humans as complex systems analysis."

That prediction proved a bit too optimistic. Whole genome sequences in 2016 still cost about $1,500, too high to become ubiquitous. The US health-care system was still struggling to implement basic electronic medical records in many cases. Even with seamless medical record keeping, not enough was known about the genome to put it onto everyone's credit card or make it part of everyday medical practice.

Radical as the prediction may have sounded in 1996, plenty of genomics scientists at the time expected that sort of rapid progress, according to a review of the literature by the *Times*. "The 21st century is going to be the century of biology," Hood said. "There's no question about it."

So many labs were using automated DNA sequencers by the mid-1990s that Applied Biosystems was raking in more than $100 million a year in sales. Caltech's royalty income, from licensing the sequencers to ABI, topped $2 million a year. With wide adoption of DNA sequencing, scientists everywhere could start asking new questions about genetic abnormalities that might be causing disease. The information gleaned from the machines was thrilling.

Scientists sequenced the first complete genome of a free-living organism, a bacterium called *Haemophilus influenzae*, once thought to cause the flu. Doing so depended on advances in "shotgun" sequencing technology, used by scientists since the early 1980s to assemble genomes. Researchers would start by shredding the genome into small segments that could be fed into a sequencer. Once the order of DNA was determined in those fragments, the overlapping ends of the DNA segments were matched up and pieced together like a jigsaw puzzle. It worked for small genomes composed of just a few thousand nucleotides, the individual letters in DNA.

But Hood and others were hunting bigger game. Sequencing was somewhat limited at the time by the need to first "map," or properly plot a given stretch of DNA at the right place on a chromosome, and then do several redundancy runs to make sure they had it right. At Hood's UW lab, Jared Roach, Cecilie Boysen, and Kai Wang wanted to see whether they could eliminate the need for the prior mapping and reduce the need for final redundancy runs. They showed that a technique called "pairwise end sequencing" could be used to assemble an entire genome when they varied the sizes of DNA fragments. The technique broke DNA into random, overlapping fragments. A computer algorithm then assembled the whole genome by matching up the overlapping segments on both ends of the fragments, instead of just one. Researchers dubbed it "double-barrel shotgun" sequencing.

"It was about figuring things out by smashing them into a million things on the floor and using software to reassemble the pieces," said Myhrvold, the former Microsoft chief technology officer, "rather than the much more careful techniques people had championed before that." Not everyone trusted this process. How would you know if the algorithm made a mistake?

But Hood had confidence in it partly because of his interactions with software experts at Microsoft. Craig Venter's institute used the method to sequence the *Haemophilus influenzae* bacterium. The institute's paper about the achievement in the July 1995 issue of *Science* was a blockbuster, cited by other researchers more than five thousand times.

It was a heady time for the UW team, full of possibilities. Lasting friendships were forged there. Young parents helped each other with babysitting. Looking back, some of the young scientists said, the first two years were like a honeymoon.

Never mind a few inconveniences.

For one, the University of Washington Medical Center was nobody's idea of a modern research building. The mazelike structure had grown in fits and starts over the years, with wings named after letters of the alphabet. (Hood would get the K wing.) It grew to accommodate not only doctors and patients but also various researchers' quirks. Storage rooms were oddly placed. Tunnels and doorways seemingly led nowhere. "We used to joke that in health sciences, we could simplify [admissions for] prospective graduate students by putting them down in the middle of the building," Lee Huntsman said. "If they could make it out, they were qualified."

Hood wasn't thrilled with the accommodations, which included crowding into part of an engineering building on the east side of campus for the first two years while waiting for renovations to the K wing. It wasn't a deal breaker; however, Hood soon realized that the entirely new building he was promised, dubbed K-2, wasn't going to happen anytime soon. UW President Richard McCormick told him to get in line behind ten other worthy applicants. Hood stewed.

Then there was the fact that UW, as a state institution, had much stricter rules and operating procedures than a private institution like Caltech. Standard protocols governed everything, down to how to order basic lab supplies. At first, Hood operated much as he had before, playing by his own rules. If he wanted to buy a piece of equipment, why not go out and raise some money, order it, and get it delivered right away? Why seek permission, or fill out a form, or wait for a committee that meets once a month? Entrepreneurs may sympathize. But at a sprawling institution accountable to taxpayers, Hood couldn't get away with his freewheeling style. He would need to learn to play the inside game, form alliances, and above all, be patient. Those who knew him well knew one thing: he was doomed from the start.

Maynard Olson saw what was happening. A classic academic, he was comfortable in this realm of bureaucracy. He was more diplomatic and politically savvy than Hood. Unlike Hood, who dismissed faculty senate meetings as a petty waste of time, Olson paid close attention to the academic environment. Olson knew how to pull the right levers to get resources for his science. In the 1970s, Olson had spent five years at UW as a "glorified postdoc" in Benjamin Hall's lab. During Olson's stint as a young scientist at UW, he switched his focus from inorganic chemistry to biology. He also learned who was who on campus. He observed. He built alliances. He learned to appreciate the customs, like waiting your turn, even when it might be irritating. He went through proper channels, paid close attention to the rules, and followed them. He liked others to do the same. Later, when his star was on the rise as an early mover in genomics at Washington University in St. Louis, Olson continued those personal habits that helped him get his way. "I'm a creature of the university," Olson said. In this way, he began to refill a reservoir of political capital at UW that he had started years earlier.

Although Hood had friends in high places when he arrived, he didn't have Olson's feel for how the place truly operated in the trenches. Hood didn't have the patience to find out. He was all action, all the time. Work hard. Play hard. Envision and execute. Tradition? Hierarchy? Obstacles? Waiting your turn? Hood barely gave those things a moment's thought. He had work to do. Figuratively speaking, he would run a bulldozer straight through campus if it was necessary to get the job done.

"You didn't need to be in the room five minutes without wondering how these two would ever work together," said Joan Goverman, the immunologist Hood had recruited from Caltech. Another deputy, Lee Rowen, put it: "These two guys were like when the irresistible force meets the immovable object."

The honeymoon, Olson saw, might not last.

Hood wasn't paying attention to that. He was on to the next frontier. The question: How might all this genomic discovery connect to real-life patient care? The goal: a cure for prostate cancer. Michael Milken, the onetime junk-bond king of Wall Street, was a prostate cancer survivor. Now he was starting a charitable foundation to fund research. Milken invited Hood to the first meeting. Milken, a man of legendary intensity and impatience, tried to spearhead a "Manhattan Project" to cure prostate cancer back in May 1993, a few months after he had been diagnosed. "Our feeling was, let's go out and recruit great thinkers who might be able to help us and ask for them to work for a year or more on cancer," Milken said at the time. "We wanted to see what we could do in a few years, not a few decades."

Prostate cancer, in those years, had seen little progress. Physicians with little training in basic biology could surgically remove cancerous prostates and give some basic hormone-deprivation drugs to cut off the flow of testosterone, which they knew helped fuel tumor growth. But the drugs and treatments had unwanted side effects, like impotence, incontinence, and low libido. Not much serious molecular biology was being done to explore the underlying genetics that might be driving the growth and spread of prostate cancer. Milken wanted to go down those paths. He also wanted to remove some of the barriers and bureaucratic red tape that he saw stalling prostate cancer research. "He wanted to persuade me to play a lead role in making prostate cancer legitimate," Hood said. "He recognized it was being done by hacks, for the most part, and he wanted first-rate scientists."

Milken wanted concrete results, right away. Just like Hood. They hit it off.

Hood gave "an impressive, revolutionary presentation," at the Milken research event, according to a foundation executive. The talk was about how functional genomics could help unravel the genetic mysteries of prostate cancer. Milken, worth a reported $550 million at the time, agreed in early 1995 to put $25 million behind a group of scientific collaborators led by Hood. Hood predicted that he'd make a difference against prostate cancer within four years.

Stephen Friend, a physician-scientist from Harvard Medical School who moved around that time to the Fred Hutchinson Cancer Research Center, articulated what gave Hood such confidence.

I had been around brilliant people. The attribute
that stood out with Lee Hood was the ability to be
completely consumed by a problem and not see any
barriers. The closest thing I know to it, in sports, is the
ability to be "in the zone." You think of Tiger Woods
at his peak. ... There is this thing called confidence.
But confidence that you're going to do something is
a combination of capability and experience, meshed
with something where you're going beyond yourself in
the passion you take to it, in making it 100 percent of
your being.

Those attributes separate Lee from normal scientists.
It's that total, total dedication. Inside. Outside.
Being possessed by an idea and having 100 percent
confidence it can be done. No question. It's a key
attribute for making something occur.

If you want to know why Lee is successful, it comes
down to total faith, not an atom, not a molecule of
hesitation. ... I have not ever met a scientist who ever
had that kind of force field that Lee has about him.
Michael Milken might be the closest comparison,
as someone who has that intensity of focus.

It had all the makings of a compelling news story: a prominent
businessman, big money, a familiar disease, cutting-edge science.
The *Wall Street Journal* ran a front-page feature that put Hood's work
in the spotlight again. "I'd be shocked if we didn't do really good
things," Hood told the *Wall Street Journal*.

Behind the scenes, Hood was immediately frustrated by what he
saw as a lack of urgency in the group and "incomplete funding" from

Milken's Cap Cure organization. "I personally have been embarrassed by the time gap in bringing the consortium to full strength," Hood wrote to a member of Milken's organization in August 1995. "It is imperative that we move ahead rapidly."

By November, a public relations blitz was on. Norman Schwarzkopf, the former US Army general from the Persian Gulf War, joined the cause. "Stormin' Norman" was a prostate cancer patient. The war hero (Schwarzkopf), businessman (Milken), and scientist (Hood), appeared together on CNN's Larry King Live show in November 1995. It was the second-highest-rated episode that year. Milken, with no science background, had made himself into "arguably one of the biggest experts in the world in prostate cancer," said one physician from the renowned Dana-Farber Cancer Institute in Boston. On the show, Milken passionately repeated his goal of curing advanced prostate cancer—in a "few years, not a few decades." They appealed to prostate cancer patients to send in their tissue samples for research. Even though Hood operated inside a medical school and seemingly would have access to plenty of tumor samples, hospitals at the time—concerned with violating patient privacy—regularly tossed them out as medical waste.

More than three thousand patients and their families lined up within a month. There was just one problem. Hood and his scientific collaborators weren't equipped to handle that many samples. They scrambled to find ways to accommodate all of the patients. They struggled to stay on track with a scientific plan that could integrate the patients' medical records, their precious samples, and their contact information so that researchers could follow up with them and see how they did over time. Adding to the confusion, one key collaborator left for a faculty job elsewhere, causing a dispute over who was responsible for the samples, and where they might end up. Hood fumed and accused a Milken associate of incompetence. Milken's group wanted Hood to map out deadlines for the delivery of scientific insights. Hood argued that science didn't work that way—it couldn't be plotted on a predictable timeline like business goals. He wrote bitter letters to the financier and other collaborators about how the project needed more money and risked imploding.

Partnerships languished in the absence of strong project management. "My failure to reach you ... leads me to conclude that you, like me, are just too overcommitted to make the Prostate Cancer Consortium a very high priority," Hood wrote to Eric Lander, then head of the Whitehead Institute, in a letter June 14, 1996. Lander responded quickly, saying his group had developed a "ferociously high-throughput genotyping facility" that might be useful if properly directed. "I haven't heard much from anyone in Seattle," Lander wrote. "I apologize if I seem uninterested. It is not the case. I am just simply unclear on what we can do to help."

While Hood told Lander privately that he was overextended, he knew better than to give that impression to donors. "This Consortium has made remarkable progress over the past couple of years, setting in place an integrated, inter-disciplinary program that is unrivaled world-wide," Hood wrote in an October 1997 letter to one supporter, an Oracle executive. "Your support has catalyzed one of the most exciting scientific endeavors of my career."

Weeks before that upbeat status report, Hood's e-mail in-box was full of complaints about the prostate cancer project. A Milken aide demanded that labs run twenty-four hours a day, seven days a week. If the lab couldn't handle all the genomic data, it should find an industrial partner to help. Hood's scientific collaborators were horrified. They thought that bringing in an industrial partner could mean not only a loss of control but also potentially a loss of the data necessary to write the papers that scientists need to advance their careers. "I'm enraged!" wrote one scientific collaborator in an e-mail to Hood. "There is no way in hell I'm sharing any part of this project with any other lab or company."

Hood had no idea how to mediate such concerns between his scientific colleagues and his wealthy business friends. The chief operating officer of Cap Cure, a leading prostate cancer physician named Richard N. Atkins, was not pleased. "For about fifteen months, I believe we have been progressively derailed from the track of progress and mutual benefit," Atkins wrote to Hood in May 1998. "Personally, I think it's a shame."

Three years in, the consortium had completed a few peer-reviewed papers and little more. The grand four-year promise from that *Wall Street Journal* article was unfulfilled.

No cure was in sight. By 1998, Hood's prostate cancer work with Michael Milken was done. The collaboration essentially ran its course, fizzled out. Hood moved on. Another scientist might have kept at it, investing his whole career. Hood didn't specialize in that way. He saw his role differently: as thinking big to alter the course of human events. He could move on to the next scientific goal without a moment's hesitation, as long as he was doing high-impact work that extended further into the future.

One way Hood planned to make an impact beyond the lab was to inspire young people. He had bold plans for revamping the way science was taught to high school students.

Science education had always been a cause close to Hood's heart, but he poured more energy into it in Seattle. Part of this was practical. Education reforms were more likely to get traction in a smaller city like Seattle than in a sprawling metropolis like Los Angeles. Part of this was personal. Valerie made it her cause, too. She devoted the final fifteen years of her career to science education, after her own kids had left home. With Hood's foresight and tenacity, combined with Logan's organized attention to detail, they made a sincere and effective team.

Could high school students sequence DNA? Hood and Logan thought so. They went to work on making science education more like real science. Hood believed that the way most science was taught discouraged children from using their imaginations, the thing they needed most to think like scientists. Hood and Logan thought kids should be encouraged to ask questions, not just memorize terms from a book. Students should have access to real tools of science. That way, they could learn by doing, rolling up their sleeves. High schoolers would learn to think and come to know the thrill of discovery.

This was a daring idea. Such an inquiry-based approach to science wasn't designed to boost scores on the standardized tests that policy makers used to measure progress. One early project in particular—the idea of getting high school students to do real DNA sequencing—sounded so unsafe as to be absurd. DNA sequencers of the early 1990s still used radiation. Insistent that students get hands-on experience, Hood dug out an older, radiation-free technology.

His goals were twofold. Hood wanted to develop a new generation of scientists, and he wanted to create a more science-literate society. If scientists explained the basic concepts of what they did to average citizens, Hood believed, then average citizens would be more likely to let their tax dollars support it. Deep down, a part of him also wanted to simply give back. Hood never forgot how his devoted science teacher, Cliff Olson back in Montana, altered the trajectory of his life.

Hood was fixed on the goal but flexible on the means. "Unlike many of his colleagues in science, Lee respected the views of science educators and was willing to listen and learn from us," said Jane Butler Kahle, a former National Science Foundation official who spearheaded science education reform in Ohio in the 1990s. "Many leading scientists at that time were concerned about the quality of science education but thought that they had all the answers."

Some scientists sought to do some community outreach just to fulfill some basic requirements attached to their federal research grants. Hood and Logan weren't satisfied with just going through the motions. They dove in. They formed partnerships with teachers and school districts to see what was needed. When Hood heard that teachers lacked high-quality training, he and Logan brokered meetings for them with scientists. He and Logan spent many evenings at school board meetings, hashing out agreements with school districts and teachers' unions. Few other leading scientists brought such passion and staying power to the issue.

The timing was good. The competitiveness of US students in a globalizing economy was a national issue in the 1996 presidential election. Hood gave stirring speeches to raise support for the cause. "One cannot underestimate the impact of a pure science institute and a scientist of Lee's stature addressing science education," Kahle said.

Hood already had financial support from the National Science Foundation for work with kids in high school. But why stop there? Why shouldn't younger pupils, those in kindergarten through fifth grade, get early exposure to science? At the time, they got little to no science instruction. In 1995, Hood and his colleagues sought a five-year, $5 million grant from the National Science Foundation to revamp science education for grade school children. The local newspapers raved. The program was ambitious. All elementary science teachers—not just the eager volunteers—would get professional training. "This way, they learn jointly and reinforce each other," Hood explained. Seattle would be a model for the nation.

One of the local newspapers spoke of reforms to science education with urgency. "Our kids need the training to succeed as individuals and we, as a nation, need them to succeed in order to stay competitive," the *Seattle Times* science reporters said. The newspaper editors went on to talk about Hood and Logan's efforts:

The Seattle scientist who has been most energetic and visible in reaching out to children is one who seems to have the least need to: UW molecular biology chairman Leroy Hood. ... He has already won a host of prizes and his own children are grown. Why do he and his wife, Valerie Logan, who administers his education effort, care what is happening with kindergartners in South Seattle? Hood is a Montana kid whose science teachers, he said, changed his life. He is deeply concerned that public ignorance of biology will collide with rapid changes in genetic engineering in the 21st century.

Hood's missionary zeal for the subject was a recurring theme in the story:

This September, new UW President Richard McCormick and Seattle School Superintendent John Stanford officially started work the same day, and both soon embraced Hood's science-education plan. "I, too, was bombarded by Lee Hood," McCormick recently recalled with humor, "in my face, saying we can do this ... and I'm saying, what, excuse me, hello?"

The newspaper also implied that Hood single-handedly started science-education programs while at Caltech and at UW. Two days later, the newspaper ran a correction:

The Science for Early Education Development Program in Pasadena, Calif., discussed in this story about science education Sunday was started

by California Institute of Technology professors
Jim Bower and Jerry Pine. The story incorrectly
stated that the program was started by Leroy
Hood, a University of Washington scientist. ...
Additionally, members of Hood's Department of
Molecular Biotechnology at the UW were not fully
credited with initiating some of the department's
education outreach ideas: Roger Bumgarner started
the elementary outreach program, and Deborah
Nickerson led the team that received an education
grant from Howard Hughes Medical Institute.

Once again, Hood had cast himself as the star of a one-man
show. What should have been a feel-good moment for everyone in
the lab became a source of resentment.

Everyone could be proud of their efforts, though: by the end
of the five-year program, more than eleven hundred teachers had
received training. That affected about twenty-three thousand students in Seattle's seventy-two elementary schools. Impressed with
Hood's tireless promotion of the cause, corporate donors such as
Boeing continued to support the program after the federal grant ran
out.

With their success, Hood and Logan looked next to middle
schools. They secured a new $1.6 million federal grant to start with
five school districts, leveraged that to get more corporate donations,
and later expanded their work to eighteen of the twenty-one school
districts in the county. They trained 350 teachers who connected
with twenty thousand students.

Results varied from district to district. When controlling for
various factors such as socioeconomic status, middle school children in the inquiry-based science program had a modest, but statistically significant, improvement in meeting state proficiency

standards from 2004 to 2010. It was hard to say that the inquiry-based program made the difference. Other schools with similar demographics that didn't participate in the program also improved. Even so, many districts in the state swore by the science education work of Hood and Logan, continuing to send their teachers and administrators in for science training sessions for many years. Even when Hood's institute ran into financial difficulties in later years, he never eliminated the science education work.

Hood's experience in the late 1990s on prostate cancer reinforced something for him. Diseases like prostate cancer that involve multiple genes, combined with multiple behavioral and environmental factors, were humbling. Hood saw the leaps in DNA instrumentation and, thanks to his friends at Microsoft, the leaps in computing power. It was time, he decided, to think more ambitiously about the greater complexity at work in the body. To study how multiple genes, multiple proteins, operated in concert. To take a holistic view.

Systems biology was the next frontier.

The concept wasn't new. In the nineteenth century, Darwin thought about evolution in terms of whole organisms. In the 1960s, when Hood was a Caltech student and faculty member trying to solve the antibody diversity riddle, scientists wondered how to even begin to approach the complexity of biology. Max Delbrück, the influential Caltech biologist, was skeptical that the deep problems of immunology could be solved one gene or one protein at a time. Looking at one gene at a time, scientists would never be able to understand how the immune system knows to fight invading cells but avoid similar structures on healthy cells, or what causes the immune system to go haywire and attack healthy tissues, as it does in rheumatoid arthritis and other autoimmune diseases. Other scientists had notions about connecting the disparate parts of biology. This emerging view was variously called systems biology, quantitative systems biology, or network biology. What Hood did in the late 1990s was create coherence around the rather fuzzy concept. He popularized the term "systems biology" and turned it into a discipline.

Hood believed the time was right for an all-out assault on systems biology. He outlined his ideas in a grant proposal to the National Science Foundation in 1998.

> The future of biology is the analysis of complex networks and systems, such as the ensemble of 100,000 genes [later discovered to be closer to 20,000] that are the blueprint for human beings, or the interactive assemblies of protein species that regulate transcription and development. To analyze and understand these complex networks and systems, one must identify the component elements, establish their interconnectivity, model their behavior computationally, and finally test the models against biological reality. The mission of the Science and Technology Center for Molecular Biotechnology [at UW] is the development of novel and powerful tools that will enable biologists to undertake such studies.

The complexity was hard to fathom. Huge stretches of genomic terrain were left to decipher. Relationships between gene structure and function would take decades to establish. Billions of DNA nucleotides give rise to trillions of proteins and trillions of cells in infinite combinations. Two decades after Hood began advocating for systems biology, scientists still didn't have 3-D predictive computer models for how a cell might respond to certain stimuli. The immune system alone was almost impossibly complex, as genes constantly rearrange to adapt to the changing real-time environment. Hood was saying biology had too narrow a focus. But biology *without* a narrow focus struck many as a fool's errand. Not so for Hood. "He has deep faith in the merits of his own science and technology vision," said Daniel Kevles, the Yale science historian. "That's what enabled him to move over to systems biology. When he moved into it, it was kind of a dicey field."

Hood also understood on a personal level the promise of gaining such knowledge about the human body.

Few people would have expected it from someone with such energy, but Hood was at risk of a heart attack. His father suffered three heart attacks, starting in his fifties, before one finally killed him at age sixty-nine. Hood's grandfather and great-grandfather both died of heart attacks.

In the early 1990s, in his midfifties, Hood's physician looked at Hood's cholesterol count and started him on a drug to lower it. After modest success with the early drugs known as statins, Hood switched to a new version called atorvastatin (Lipitor). His doctor prescribed it because the drug was generally effective for a population of similar patients. Why not try it and see what happened? The drug, which became the biggest-selling pharmaceutical product of all time, worked for millions.

Hood took the drug for a few years, but eventually developed severe muscle pain. "It was terrible for me," Hood said. No one knew for sure why the drug worked for so many people, but caused significant muscle pain for some.

Hood switched to another drug in the class, simvastatin (Zocor), and it worked better. No one knew why. The biologist in him wondered: Shouldn't someone be able to analyze his disease at a molecular level to see how it differed from another person's, and match up the underlying problem with a therapy that was highly likely to work? Prescriptions would then be given on a more rational, individual, predictive basis. The idea had been around a long time. The Human Genome Project was sold on the promise that it would deliver "personalized medicine." Hood wanted to see medicine move this way for the good of biomedical science, but also for himself and his family.

Systems biology was the way forward. Even if others didn't see it.

Hood's new quest was not a radical departure from the original vision for the UW Department of Molecular Biotechnology, but rather an evolution of it. Hood still needed an interdisciplinary team of instrument people, biology people, chemistry people, and computer people.

One of those computer people was Trey Ideker. With undergraduate and master's degrees in electrical engineering and computer science from MIT, he was a quantitative thinker. The messy ambiguity of wet lab biology was foreign to him. Ideker had none of the usual prerequisites—like organic chemistry, genetics, and developmental biology—that would typically be expected of a graduate student applying to a molecular biology department. But Hood needed all types, and that meant being flexible about which kinds of students to admit. Ideker was restless. He didn't want to work at a defense contractor "making bombs." He read about Hood and the Gates-funded department. He made his pitch to join the first incoming graduate students in 1995.

Within a couple years, Ideker, a clean-cut, earnest type, was Hood's favorite student. For Ideker's part, "I was completely spellbound," he said. "Lee is like the Bill Clinton of biology. He's got this crazy blend of charisma and intelligence." By the late '90s, Ideker was working directly with the department head on his most exciting projects. He would go on to join the faculty at the University of California, San Diego. But before that, Ideker developed a program called Cytoscape, which helped visualize complex biological network data. It was one of the "novel and powerful tools" Hood envisioned for advancing systems biology.

He imagined his UW department studying systems biology for years to come. There was, however, a problem. Hood was wearing out his welcome. On one hand, his foresight, drive, and international reputation were assets for a university that was hungry to make a name for itself on the global scientific stage. But Hood's sense of entitlement and willingness to break rules were frustrating UW officials, much as they did at Caltech. Administrative duties still weren't Hood's strong suit. Many university chairs see their role as that of rainmaker and remover of obstacles for fellow faculty. Hood sometimes behaved more like a start-up CEO, deputizing faculty to work on his projects like underlings, rather than treating them as peers, as heads of their own academic kingdoms. Questions over people's roles, career tracks, and proper status usually fell on Hood's deaf ears. Resentments simmered.

The *Wall Street Journal* published a story about Hood's UW department in May 1995. It was a golden piece of publicity that would open doors for Hood among powerful businesspeople for years. But it also noted a culture clash between the technology and biology sides of the department. One assistant called the lab a place of "creative anarchy." Alan Blanchard, the Caltech recruit who repurposed the ink-jet printers, told the newspaper that Hood knew the biology but not the technology. "I've gotten no direction from Hood," Blanchard said. "Actually, the only direction he ever gave me was wrong."

The department gained a reputation as dysfunctional. Alan Aderem, an immunologist from Rockefeller University, was drawn to work with Hood but was careful to keep a certain distance. "I wasn't going to go into a department with Lee," Aderem said. "I wanted to go in as an equal in leading a new initiative. It had nothing to do with ego. It had everything to do with being able to influence the agenda sufficiently and not be run down by Lee. I knew, from everything I knew about Lee, that his enthusiasm will drive him in a direction—and if I didn't think it was the right way to go—it would be a real problem."

As he had at Caltech, Hood avoided certain confrontations. He found it extremely difficult to fire people, for example. That meant people could hang around for years, publish next to nothing of scientific significance, and never face any consequences from the boss.

"Lee is like the sun," observed Perlmutter. "He's this sun that throws off tremendous energy. And there are a lot of moons that orbit him, that don't throw off much energy themselves, but they reflect it. You get too many moons, and it's not so good."

Hood had defenders. His most powerful ally was Fialkow, the physician-scientist who ran the UW School of Medicine. When others complained about Hood's loose-cannon ways, Fialkow would often calm them down and say he'd straighten things out. It helped that even the harshest critics believed that Hood, while often wrong and reckless in administration, had good intentions. No one thought Hood was abusive, vindictive, or Machiavellian, like some scientists of his caliber. Still, the balance of opinion began to tip.

Just as Hood's clout on campus was fading, in November 1996, tragedy struck. Fialkow, at age sixty-two, was killed in a storm during a trek in the Himalayas. Suddenly, Hood had lost his most important heat shield. The honeymoon at UW was over.

Chapter 14

A Messy Divorce

Molecular Dynamics was thrilled. The esteemed Lee Hood had just purchased one of the start-up company's DNA microarray scanners, which measure gene expression in biological samples. On the eve of an important investor conference in January 1998, Molecular Dynamics issued a press release touting this external validation of its technology. Hood also apparently agreed to help the young company further develop its technology.

It was a publicity coup for a young company seeking to woo new customers and investors. "The University of Washington has the expertise needed to refine and push the applications of our technology, which will ultimately benefit our other program partners as well as future system users," said Molecular Dynamics CEO Jay Flatley. A shrewd businessman, Flatley later left to run Illumina, which became the dominant DNA sequencing company of the early twenty-first century.

Hood hadn't been very shrewd. A competitor, Rosetta Inpharmatics, saw the news and hit the roof. Hood had cofounded Rosetta the previous year. He'd licensed Alan Blanchard's ink-jet oligonucleotide printing technology out of his UW lab to create the start-up, which focused on computational analysis of gene expression data. "Rosetta was completely blindsided by the release and dismayed to find one of its founders in bed with a competitor," an internal memo said. Rosetta asked Hood to resign from its scientific advisory board. Hood had made the purchase over Christmas break, when most everyone else at UW was on vacation.

Within a couple of years at the university, Hood was creating chaos in his administrative role as chair of the Department of Molecular Biotechnology.

Accounting was a nightmare. To pay department bills, Hood tended to reach into one of his two "discretionary" accounts—the Gates funding, or his Caltech royalty stream. It was nearly impossible to keep track of where all the money came from, what it was spent on, and how everything was commingled. An internal audit in 1994 found "systems weaknesses that could cause unallowable expenses to be charged to sponsored agreements." Yet during university budget cuts, Hood asked administrators for special treatment. "Cuts should be focused on eliminating mediocrity and not on cutting everybody, good and bad alike," Hood wrote to the UW dean of medicine.

Impatient over securing lab space for an ambitious plant-genome sequencing project, Hood hired his own agent to negotiate over a particular space. It turned out the university also was negotiating for that space—for Hood. Essentially Hood, in his haste, created a bidding war against himself. Previously, he had berated a university real-estate official for dragging her feet. "I got a call from Lee

Hood today, who is 'very' angry that we haven't found lab space for him yet," the real-estate official wrote in a memo to her boss. "He demanded that the UW live up to its commitment to him. ... I could barely get a word in edgewise. ... I explained that there's a shortage, and it's highly competitive. ... Dr. Hood's anger at me, and the university, for 'doing nothing' was unwarranted."

The state passed an ethics law in 1994 to prevent public employees from using public resources for personal gain. Hood complained that it would virtually shut down entrepreneurship among faculty members. "A UW professor would have to resign before accepting the Nobel Prize," Hood exaggerated.

Inside his university department, Hood clashed with Olson early and often over technical and strategic questions. Above all, their dramatic clash of personalities loomed. One man radiated frenetic energy and sowed chaos; the other was calm and craved order. Olson couldn't stand the lack of meticulous financials, records, and experimental data. He was almost "morally offended" by Hood's style, fellow scientist Lee Rowan said. Olson was a classic academic scientist. He didn't care for the way Hood spun wonderful tales for financiers, or for his brand of entrepreneurial science that led to so much overpromising. "The more people knew about the particulars of what [Hood] was talking about [in genomics], the less tolerant they were of the way he turned it into an almost political-style message," Olson said. "It's a message that's evangelical and brings in converts. The Lee Hood message wasn't moored in the messy realities of experimental biology. It kind of floated above that."

Time and again, scientific reviewers tore Hood's lofty grant proposals to shreds. One proposal for a three-year NIH grant was audacious: sequencing five megabases of DNA in the first year, twenty-five in the second, and seventy-five in the third. At the time of the application, the Hood lab had completed only two or three megabases. "The NIH totally shit-canned that grant," Rowen said.

"But they wrote a lot of thoughtful, helpful criticisms. NIH came back to us and said, how about two megabases, five megabases, and eight megabases? Even I thought we could do more. We wrote the grant for three, six, and twelve. They gave it to us." A reviewer from the National Science Foundation was skeptical of the claims in another proposal: "It is impressive to say that a new low-pass sequencing strategy will be developed with a 100x cost/time benefit over current methods, but nothing is added to even hint at how this claim might be realized, or how it improves on the prior low-pass gene skimming strategies. The rest of this section continues in the same 'trust us' vein."

Sloppiness also cost Hood in his first major grant proposal for a federally funded UW Genome Center. In 1995, after a site visit by the National Institutes of Health, one reviewer wrote: "When asking for $67 million over five years, one would rather like to see some convincing reasons to spend it." The reviewer criticized the grant proposal as a collection of "nearly unconnected, separate projects" with "no sense of compelling reasons to suggest why the proposed methods are the best way to proceed."

Hood was advocating—both in the grant proposal and in public—ways to speed up gene sequencing. To enable his brand of high-speed shotgun sequencing, Hood embraced the use of bacterial artificial chromosomes, developed by Melvin Simon's lab at Caltech. Short pieces of human DNA could be put in the BAC—an artificial DNA construct placed in a bacterial cell. BACs were handy because they could be used to amplify the short pieces of human DNA before they were fed into the sequencer. Another advantage: BACs could be used to build long cloned stretches of DNA, fed into the machines simultaneously. By using this technique, researchers could proceed across long, repetitive regions of DNA and still be confident that they were aligning the correct matching ends. Other cloning methods used to prepare samples worked only with short stretches of DNA. These sometimes failed to remain stable during a run through the sequencing machine. Longer term, Hood hoped the BAC method could also allow researchers to skip the tedious work of "mapping" DNA. In the early years of the genome project, researchers had to first line up stretches of DNA into their proper position on chromosomes, then sequence from that point of reference. But if scientists could feed longer stretches of DNA into the machines and properly align overlapping ends of sequences, they could assemble a whole genome faster.

The BAC method was Hood's style: big, bold, and fast for its day. More conservative scientists, including Olson, preferred to stick with the tried and true. Olson had previously developed a technique that used yeast artificial chromosomes (YACs) for cloning segments of DNA before sequencing. These DNA constructs, which operated in yeast cells, could also do the job of preparing many copies of human DNA segments for sequencing. The method was meticulous, proven, and established.

This new BAC method at first seemed sloppy to Olson. What good was a genome if it wasn't accurate? Hood shot back: What good was a genome if it took forever to get it? When it came to the Human Genome Project, promises had been made to Congress and the public about delivering the genome on budget and on time. The technology needed to improve in a hurry. That meant going out on a limb, tweaking prototypes, trying new methods. Olson, who knew the technical aspects of sequencing inside out, appreciated the practical argument. He'd written an influential editorial in *Science* in October 1995 titled "A Time to Sequence." ("For Hood, it was always a time to sequence," Olson said with a chuckle, years later.) Olson acknowledged some of the technical debates, saying there was a "healthy competition" between those advocating BACs and those who preferred another system. It's never easy to design an airplane while flying it at the same time.

But as the NIH grant reviewers visited Hood's lab to consider his 1995 proposal for a genome center, they saw too much disarray. They also knew of the simmering feud between Hood and Olson. "It is a bit depressing that such an impressive collection of bright and talented people could not coalesce better," one reviewer wrote. "If this is the best that can come out of an academic group that wants to be funded as a genome center, it is no wonder that the NIH-funded genome centers have performed so poorly."

They rejected the initial grant proposal.

Two years after joining Hood with so much fanfare and excitement, Olson began to distance himself. In 1994, he stopped sharing financial information about his grant proposals with the department chair. Characteristically, Hood didn't confront him about it. Other administrators complained about Olson's secrecy and Hood's unwillingness to address conflict.

By the summer of 1996, Olson was at the end of his rope. He approached Paul Ramsey, then an assistant dean of the medical school, about transferring to another department. Olson offered to surrender some prime lab space in the Health Sciences complex and carry on in Fluke Hall—across campus from Hood's group. Olson wanted a clean break. He didn't want to see the man anymore. In a memo to university administrators, Olson wrote:

> It is obvious that Hood's activities and mine will be most harmonious if we restructure our interactions along the principle that "good fences make good neighbors." ... The current situation is alarmingly unstable. New problems arise almost weekly.
>
> The sooner and more decisively we move from ad hoc management of these problems ... the better the chance of long-term success.

"I was a [principal investigator] with grants worth millions of dollars a year," Olson said years later. "When you're operating at that scale, you can't deal with uncertain administrative arrangements." Hood, for his part, remembered personality as the main problem, more than managerial strife. "Maynard ... was interested in doing science in his own way," Hood said. "Maynard is really careful, and calculating and precise, and I'm much more flexible and loose and give people a lot more freedom. He didn't think that was a good idea."

Olson tried recruiting away some of Hood's postdocs and technicians, promising sanity and order. "You had the Olson camp and the Hood camp," said Todd Smith, a former postdoc who initially sided with Olson but switched back to Hood. "Most departments don't have anyone who ever talks to Congress. We had two people. And they disagreed."

Hood looked to his most powerful behind-the-scenes fixer, the medical school dean Phil Fialkow, to help. "It is important Maynard be told clearly his behavior is strikingly inappropriate and that it should cease," Hood wrote to Fialkow. Just a couple months later, in November 1996, Fialkow died. The new dean of medicine, Paul Ramsey, hadn't been part of the Hood welcoming committee. He hadn't made Hood any promises. Olson's transfer was completed that same month.

As the department splintered, many people felt torn. About Hood, they tended to feel both intense repulsion and attraction. There was the chaotic management style, but also the ideas, enthusiasm, and manic energy that continued to pour out of the man. Without a doubt, he inspired many to do their best creative work. People wanted to be a part of that. For example: Graduate students and postdocs typically stay in a lab for a few years. They try to get at least a couple of solid published papers to their name, and then they leave to establish their own identities as independent researchers. But Hood's force of personality—and stream of exciting, fully funded projects—was magnetic, especially for people who didn't relish the prospect of raising their own money. Tim Hunkapiller stayed with him almost twenty years. Lee Rowen stayed longer.

Todd Smith was one who couldn't stay away for long. When he joined Hood's lab, he had a PhD in medicinal chemistry from the University of Washington, plus five years of biotech industry experience. He spearheaded a collaboration with UW geneticist Mary-Claire King in which they sequenced the BRCA1 gene. King was already famous for identifying the gene and its hereditary role in increasing a woman's risk of breast and ovarian cancer when mutated. Smith was excited about the high-profile project. He also was interested in computers, using software to analyze DNA. He had a desirable set of skills for the department Hood was building.

But after three years, in spring 1997, Smith was weary of the chaos. He joined a group of four disillusioned postdocs who sided with Olson. Then he got a job at a start-up backed by Kleiner Perkins Caufield & Byers, the venture capital firm. Boldly, Smith took two UW colleagues with him. "Hood was mad that I left," Smith said, "and even madder that I took people with me."

One month later, Smith realized joining the start-up was a mistake. He started over with his own company, called Geospiza. Hood wasn't one to hold a grudge or say "I told you so." Instead, he hired Smith's new company to do $65,000 worth of work on a genetic-data visualization project for the National Science Foundation. It was a small contract, but it was a generous move that provided validation for Smith's start-up. It helped Smith raise further investment.

As he learned to become a manager, Smith saw the good and bad in Hood's freewheeling style. "He's the shark in water; he keeps moving forward," Smith said. "A lot of leaders have this, that really central drive. All he wants to do is science. That singular passion, and focus, is his defining characteristic. It also leads to all the turmoil that goes on all around."

The rejected genome-center grant and the split with Olson weren't going to keep Hood from moving forward. He hunted for a new angle in genomics. Drawing on his background in immunology, Hood rebounded from the initial NIH grant rejection and won support for a niche genome sequencing project on chromosomes 14 and 15, which together represent 6–7 percent of the human genome. Lee Rowen was overseeing a small group focused on sequencing T cell receptors, the major histocompatibility complex (MHC) of the immune system, and long regions of chromosome 14. Those were interesting projects with legitimate medical interest because MHC was involved in tissue compatibility for human tissue transplants. T cell receptors were important to adaptive immunity—the body's ability to fight off infections. But these were still fairly small niche projects in the grand scheme of sequencing.

Hood had his eye on a bigger prize, something that would make a splash. Something that could compete with a new big name in science: Craig Venter.

The biggest promise of DNA sequencing—completing the human genome—inched along. For Venter, a daring entrepreneur and former NIH scientist, it was too slow. The federally funded project, he said, was bogged down by bureaucracy. The NIH's own grant reviewers said as much. Venter had quit the NIH in 1992 to start his own institute, insisting there were better ways to approach the Human Genome Project. By 1998, he took charge of a company to prove it: Celera. He claimed he could sequence the entire human genome in half the time—at a tenth of the cost—with private money instead of public money.

Those were fighting words.

Generations of scientists had been supported by federal funding. Classic academics, pursuing fame and Nobel Prizes, didn't appreciate the implication that their work was boondoggle. Researchers were peeved that someone would try biting off the hand that fed them. In Congress, the political parties had set aside their rancor to support biomedical research. US policy makers were searching for advantages in a globalizing economy, and they saw a domestic source of strength in biotech. As the Human Genome Project gained momentum, a deal was struck: Republican House Speaker Newt Gingrich and Democratic President Bill Clinton agreed to double the budget of the National Institutes of Health over five years, from 1997 to 2002. If private money swooped in and showed up everyone else, public support for such federal budget increases might evaporate. Yes, Hood and others had raised private money for their research, but only to make the pie bigger. They had never pitted one funding source against another like this.

Plus, private money rarely had the public's interest at heart. Profit, not knowledge, was the main motivation. In the public project, the completed genome would be free and available to all researchers. Celera spoke of keeping it private and selling access to pharmaceutical companies. This sparked a great philosophical battle.

Venter was on a crusade, and his timing was flawless. The economy was booming in the mid- to late 1990s. Millions of people logged on to the Internet for the first time. Optimistic visions for the future were everywhere in the popular media. Investors were seeking the next big thing after the dot-coms. They poured billions of dollars into aspiring genomic companies like Celera. The market for selling private DNA code was thought to be immense.

The battle lines were drawn; the competition between public and private interest was on. "Darth Venter" became the most hated man in science.

But not to Lee Hood. He could relate to Venter in some ways. He knew what it was like to imagine a better way of doing things. To pursue an idea with single-minded tenacity. To be a showman. To be skilled at raising money. To relish proving the critics wrong. Maynard Olson acidly observed years later: "Lee Hood is kind of like

Craig Venter, but with real accomplishments. Craig Venter skipped over the accomplishment phase of his career. I wouldn't accuse Lee of that."

Like Venter, Hood in the mid-1990s was increasingly impatient with the cautious, incremental nature of the federal funding system. Hood sometimes sided with Venter in technical debates over how best to kick-start the genome project. After reading about Venter's 1995 triumph with sequencing *Haemophilus influenzae*, Hood urged one of his scientists, Greg Mahairas, to keep improving the BAC technique—as applied to plants. In collaboration with Venter, Hood scraped together funding from the Department of Energy (which had a long-standing interest in DNA sequencing) for a pilot project. Hood, Venter, and Hamilton Smith coauthored a May 1996 paper in *Nature* advocating large-scale shotgun sequencing as a way forward in human genome sequencing.

Mainly, though, Hood saw Venter as a competitor to be beaten.

Hood wasn't trying to race Venter directly on the Human Genome Project. While he paid attention to it, and oversaw a small UW genome center that sequenced parts of chromosomes 14 and 15, Hood never got involved in high-level day-to-day management of the international consortium. That work was largely done by Francis Collins, Eric Lander, John Sulston, and Robert Waterston. Hood remained mostly on the periphery, thinking about strategies for sequencing and assembly, as well as relentlessly pushing new technology development. Sequencing technology was only one part of his grand plan for interdisciplinary biology at UW, which included proteomics, gene expression, cell sorting, and bioinformatics. He had plates spinning in all of these areas.

Still, he wanted to make a splash in genomics. At the time, the federal government had concentrated much its funding on centers at Washington University in St. Louis; the Whitehead Institute in Cambridge, Massachusetts; and Baylor College of Medicine in

Houston. But there was more work to go around, sequencing not just the human genome, but other species as well. Hood had an idea: rice.

It was an odd choice for a career-long human biologist. It was an odd choice for a medical school. But the Department of Energy pilot project with Venter, focused on sequencing plant genomes, had done well. Hood secured $15 million more from the Department of Energy to see how far Mahairas could take it. Scientifically, plant genomes were interesting. And now Mahairas had ironed out some nagging technical issues with BAC sequencing. It was getting exciting. "We started thinking 'OK, we can do this on any large genome,'" Mahairas said.

Huge agricultural businesses had both money and motivation. If they knew the DNA sequences of cash crops like rice, corn, soybeans, and sorghum, it could be an important step toward engineering in desirable traits, like drought or pesticide resistance. Hood knew some people at Monsanto who wanted a piece of the genome action. Hood's ties to Monsanto dated to the early 1980s at Caltech, when the company had sponsored research on the DNA sequencer. By the late 1990s, Monsanto was interested in rice.

Rice was an enticing opportunity. Half of the world's population counted on it as a staple food. Wheat was an important crop, too, but a larger and more complex genome than rice, thought to be beyond the reach of sequencing capabilities at the time. The rice genome was both important and attainable. It was the third-largest genome project undertaken at the time, after human and mouse. Whoever got the rice genome first would make the history books.

Monsanto was ready to pounce. Hood was ready to steal thunder from Craig Venter. There was talk of a ten-year partnership. In July 1998, Hood laid out a plan to Monsanto CEO Robert Shapiro:

> There is a unique opportunity for Monsanto to
> take a major leadership role in plant genomics

by carrying out a low-pass sequence analysis of the entire rice genome. There is an enormous opportunity to garner significant intellectual property (e.g., patenting control regions as well as many potentially important undiscovered genes). If we do this, the [UW] and Monsanto will have to create an effective pipeline and analytic tools for handling enormous amounts of data. This is a major step beyond the Perkin-Elmer/Venter proposal to sequence the human genome.

An international consortium of researchers had a head start. But with the BAC method and Monsanto money together under one roof, Hood thought he could chase them down from behind. Monsanto wrote a $30 million check. It was off to the races.

Greg Mahairas had been working in a small room on the main campus—fine for small-scale experiments, but no place for an industrial-scale assault. Lab members competed for time to use the sequencers. That would never fly for a project that was bound to be a full-throttle competitive race. Hood found a facility off campus in Seattle's Queen Anne neighborhood, a twenty-minute drive from the university. In less than three years, it ballooned into the biggest operation Hood had ever managed, growing from five people to 275 at its peak. The facility brought in almost $100 million of sponsored research.

The scene was chaotic even by Hood's standards. Mahairas, the primary manager of the site, was surprised one day to see thirty-five new sequencers show up at the loading dock. About 120 such machines ran full tilt at the peak. The prize: the sequence of a variety of long-grain rice cultivated in Japan called the Nipponbare. "It was the biggest and the baddest," Mahairas said. "It was insanity. It was a huge competition."

Once, when Mahairas was at the office about 1:00 a.m., he heard a knock. There was Hood, tapping on his window. Hood didn't have a key to his own building. Along with Hood in the middle of the night were a couple of Nobel Laureates—Phil Sharp of MIT and David Baltimore of Caltech. They wanted a tour.

By 1999, DNA sequencers were no longer Model A Fords. Applied Biosystems had figured out how to build sequencers that handled samples in parallel, through capillary tubes, instead of the previous slab-gel method. "High-throughput" sequencing was like trading in your dial-up Internet modem for fiber-optic cable. Technicians couldn't keep up with the data coming from the machines. Hood's lab set up a virtual private network to ship DNA data every night to Monsanto headquarters in St. Louis.

Sequencers ran twenty-four hours a day, seven days a week. Hood assured his financial backers that his team could hit aggressive timelines. So when something broke, it needed to be fixed immediately. The frenetic pace also led to carelessness in handling samples. At one point, glass slides got scratched, which led to botched sequencing runs, wasting $50,000 each. Mahairas spent $10 million to replace all the glass at once. Just like that, a multimillion-dollar decision was made while Hood was out of town. Mahairas knew the boss would support him. The university was another matter.

The UW Board of Regents asked Hood to testify at one meeting to explain why he bought $5 million of sequencers in one fell swoop. At one point, a special department administrator was brought in to try to understand what was going on and get the rogue operation under control. "We were perceived by the whole medical school as crazy," Mahairas said.

Olson could only shake his head. "I never set foot over there," he said. "I had too much on my plate, and I could see it wasn't going to work. Anyone who knew anything about genomics realized it wasn't going to work. There were no experienced scientists."

The rice genome project flew off the rails near the end of 1999. Monsanto was under fire from European activists, upset with its work on genetically modified organisms. With the company under attack, it pulled the plug on the genomics project. Mahairas recalls being told that he had six months to wind down the operation and lay off the entire staff. Hood insisted that the project, about 80–90 percent complete, needed only about three more months of financial support to wrap up. Monsanto gave him no wiggle room. To the horror of all, Monsanto exercised its right to take all of the sequence data from the project for its proprietary purposes. That meant the researchers who had worked so hard wouldn't have access to the data needed to write papers that could advance their careers.

It was as if an entire, $100 million biotech company was born, raised, and died in the matter of two years.

"People were so upset," Mahairas said. "There were things disappearing. One night, someone goes and takes a crap right in the middle of the bathroom floor."

Hood had courted Monsanto. Hood had agreed to the insane deadlines. Hood had made promises he couldn't keep, at least not without creating an environment where $10 million got flushed in a day and people worked around the clock. Hood was ultimately in charge of the project. Yet he didn't take even half of the responsibility for its spectacular failure:

> If Monsanto hadn't had that tragic GMO challenge
> just as we were going toward completion of the rice
> genome, we would have had it done first. We had it
> 85–90 percent done. It was a crisis the likes of which
> I never saw. … They just shut us down completely.
> That was probably a year and a half's work, where we
> never got one publication out of it. Monsanto swept
> away all the data.

Olson, by this time running his own NIH-funded genome center, hired a few of the people who had been laid off. Hood said he tried to help people find other jobs, but "there was only so much I could do."

The nightmare had all started with a dream.

Hood wanted to create his own institute within the university— an institute unburdened by the incremental thinking of the federal government. He'd been thinking about it seriously since late 1996 and early 1997. If Hood raised $100 million to $150 million in private money, he could create an endowment that threw off investment earnings every year, which he could spend with no strings attached. He could more readily pursue his most audacious long-term goals. Goals like systems biology, the attempt to understand how entire biological networks operate in a symphony—way beyond what could be learned from studying just one gene or protein at a time. Nobody in the NIH peer-reviewed grant-funding system was ready to go out on a limb for that just yet. Too much mutual back scratching went on there, Hood thought, among scientists devoted to supporting their small-scale research projects.

Hood also was tired of having to answer to the dean of medicine. He never got along with Paul Ramsey. One time, shortly after Ramsey succeeded Fialkow as dean, Hood wanted to hire a surface chemist, someone expert at studying chemical reactions at surfaces and interfaces. Ramsey asked why a medical school needed a surface chemist. Hood gave a brief explanation.

"I don't buy it," Ramsey replied.

"Well, I'm going to use my Gates endowment money to hire him," Hood shot back.

"I'm not going to let you use the money to do that," Ramsey replied, and walked out.

Hood fumed. The dean had that power, but the previous dean had always given Hood latitude. Ramsey was running a tighter ship. The surface chemist went elsewhere.

Technology transfer was another source of conflict between Hood and the university higher-ups. The tech transfer office is responsible for seeking patents on university inventions, building start-up companies, or licensing the intellectual property to businesses that develop products. Hood (and Caltech) had made millions of dollars this way, dating to the formation of Amgen and Applied Biosystems.

Hood believed the UW tech transfer office was too cautious, bordering on incompetent. Afraid of being snookered into giving away a golden goose by savvy industry deal makers, inexperienced university staff swung too far in the other direction, holding out for rich licensing terms that businesspeople considered onerous. At a public university, the office also had to be attuned to political perception—avoiding any scandalous accusations that public resources were being used for personal gain. Hood wasn't the only faculty member who saw dysfunction in the way the office worked. He was just the most visible critic on campus.

If he had his own institute, Hood would be free from all that. He could raise money from whomever he wanted, bring in the exact people he wanted to carry out his mission, and license out technology on his own terms.

Institutes affiliated with universities were not unheard of. Hood studied the model established by the Whitehead Institute—an endowed, independent institute near MIT. Whitehead director Gerry Fink told Hood that an independent institute needed to be nimble, flexible, and free of oppressive micromanagement from its big academic brother. The Whitehead gave its scientists freedom to pursue their own projects. They maintained prestigious faculty positions, complete with access to students at MIT who could carry out experiments. But MIT and the Whitehead were different animals from UW. They were private institutions. They weren't accountable in the same way to state lawmakers and taxpayers.

Hood was setting himself up for a battle.

UW administrators were not gung ho about the idea. But they did make an effort to accommodate Hood, according to voluminous departmental records in the university archives. Hood had attracted a lot of talented young scientists, raised a lot of money, and brought a lot of international cachet to the university. It was hard to live with him, but they didn't want to lose him.

Hood barreled ahead with fund-raising. He had no trouble getting meetings with high-level businesspeople.

John Doerr, the venture capitalist at Kleiner Perkins Caufield & Byers, suggested meeting executives at Microsoft, Oracle software billionaire Larry Ellison, and others. Within days, Hood was on a plane to see Ellison. "It was a great visit yesterday," Hood wrote to the Oracle chief in February 1997. "Your house, as you know, is spectacular. I hope I will get to see the new one."

Like any good fund-raiser, Hood tailored his message to the audience. One e-mail to John Neilson, an early Microsoft executive, spoke of a great intellectual puzzle, a big market, and a chance to make history. He wrote:

> With Microsoft, I see the areas of collaboration
> include the need to generate databases that can
> handle extremely heterogeneous data and at the
> same time be able to generate complex cross-
> querying. This, I believe, could have important
> implications for medicine, human genetics, and
> biological database opportunities as we move into
> the twenty-first century. ... With the expertise
> Microsoft has, I suspect there would be some people
> fascinated by the challenge, of, for example, taking
> DNA sequence data upon which evolution has had
> 3.7 billion years to inscribe information, and figure
> out how to decipher that information.

Hood asked Neilson for time to make his case to Nathan Myhrvold, the Microsoft chief technology officer, and Gates. Neilson said he'd work on it.

By fall 1997, UW scientist Robert Franza was charged with working out details on the concept of a quasi-independent institute within the university.

Hood didn't like the "quasi" part. "I am skeptical about whether it is possible to establish the Institute as part of the University of Washington," Hood wrote in an October 1997 memo to Provost Lee Huntsman. "I believe the Institute needs to control its hiring, purchasing, intellectual property, endowment, etc." Hood said he was concerned that if he and his staff were still state employees, they might be hampered by the restrictive new Washington State law on ethics in public service. The dispute was well documented by an early technology for the time—electronic mail correspondence. Hood opened up his e-mail files for review, many of which he had his assistant print for posterity. Copies of many of the same e-mails, memos, and more, were available in the University of Washington archives.

After some of these meetings with Hood and his colleagues, Huntsman worked hard to control his temper.

"Lee started to go off today about the fact that it is obvious that an arrangement of the complexity of the Institute will require one to three years of University deliberation before a decision can be made," Huntsman told Dean Ramsey in an e-mail. "If the others hadn't been there, I would have sworn at him. ... I think the University has moved quickly and correctly and they ought to be bloody appreciative." He concluded the memo by telling Ramsey "an executive SWAT team" was being assembled to "discern the realm of the feasible."

University officials hoped to strike a deal. President Richard McCormick sent a handwritten note to Hood on November 5, 1997, that dangled the possibility of a scientific dream home dubbed "K-2"—an extension to the K wing where Hood worked. The name conjured the grand image of the world's second-highest peak, which appealed to the outdoorsman in Hood. McCormick said he was optimistic he could secure the funds for the new palace. Huntsman had delivered some positive progress reports to the university president. "I am pleased to hear that those discussions are going well, because your work is extremely important and valuable in so many areas," McCormick wrote.

Things weren't actually going well. Rumors spread about behind-the-scenes tensions. This time, a surprising suitor came calling. In the six years since Hood had left Caltech, David Baltimore had taken over as president. One of the world's foremost biologists, Baltimore toyed with the idea of bringing Hood back. Baltimore believed that Hood's vision for cross-disciplinary biology was farsighted and that Caltech had made a mistake by retreating into small-science fiefdoms after Hood left. Hood sketched out a rough proposal. Baltimore replied in February 1998 with cautious optimism:

> I've been thinking over the proposal you sent for your institute, and I got to wondering if there is a shot at setting you up back here. I think you will be happiest if the Institute is independent, but has an academic partnership. I'm not sure that Caltech is ready for it. ... I so often hear your name around here as the paradigm of something exciting but fearful. I think that in the past, the fear ruled, but here is where I sense a change of attitude. Perhaps they are ready for the excitement to rule.

There were allies on campus—particularly in physics, chemistry, and the humanities—who encouraged Baltimore. But not in biology. "The feeling, particularly in biology, was so strong against him that it wasn't going to work," Baltimore said years later.

Hood's files don't show any evidence that he seriously considered returning to Caltech. He kept negotiating with the University of Washington for his institute. Hood wrote the following in a June 25, 1998, memo to Paul Ramsey and Bob Gust, another administrator:

> I am concerned we are not moving forward with the Institute at a reasonable pace. It is time to change the process of taking issues one at a time, and move toward resolution of all the remaining issues by July 14 [1998]. I have several important negotiations coming up for the Institute, and I need resolution of all the major issues in order to proceed.

A week later, Gust wrote back: "I am confident we can work together."

Hood was on an emotional roller coaster. In one optimistic moment, he told Ramsey he was ready to charge ahead and make his best pitch to Bill Gates, Paul Allen, and Craig McCaw, the Seattle telecommunications billionaire. In the next moment, Hood learned that an important meeting with Gust couldn't occur because he was booked solid for a month. "It took three weeks to get a five-minute phone call," Hood complained. Hood hounded Gust and Ramsey to get on their calendars.

Early proposals from the administration revealed some large differences from what Hood had in mind. Hood imagined himself calling the shots. Administrators had other ideas. "Particularly horrifying was the initial draft of the governance document where the Dean was to have absolute control of everything," Hood wrote.

That summer of 1998, Hood tangled with administrators over matters large and small. In one humiliating move, Hood was reminded that Ramsey needed to sign off on travel expense vouchers for his tiniest cash expenses, like parking at the airport. Around that time, a friend at the Department of Energy tipped Hood off to a university clerical error that jeopardized a $2.5 million federal grant to support his genomics work. Hood exploded. "I am astounded that the University could lose track of such a large amount of money, and that the only reason it was saved from being reclaimed by the Treasury Department was that the head of the DOE Genome Program indirectly called me. ... I am frustrated that I can not get clear explanations regarding this event in a timely manner from our University bureaucracy. I am embarrassed and angry," he wrote at the time.

Ramsey had the power to keep everybody at the table, though. Hood knew the dean had gone out on a limb for him on the Monsanto project—given that Monsanto was a lightning rod for activists and that plenty of people on campus were uncomfortable with its sponsorship of academic research. Hood didn't want his boss to shoot down any existing or future agreements with Monsanto or any other potentially important corporate sponsor.

Ramsey, a cautious and deliberate administrator, lectured Hood in writing on proper procedures at a public institution. Ramsey tried to assure Hood that the dreaded "bureaucracy" was actually working to help make his vision a reality. But Ramsey's patience was wearing thin. In a five-page memo in October 1998, Ramsey wrote:

> We have been timely and generous with the School of Medicine's resources, and flexible with our response to your somewhat dynamic requirements. ... Despite the considerable commitment of the [School of Medicine] and the notable progress that has been made over the past year on this complex undertaking, we continue to receive ambiguous remarks from you. ... I have heard

that you are dissatisfied. ... I am perplexed that you
believe our unprecedented efforts are inadequate.

Ramsey tried to leave each side room to maneuver: "I am anxious to move together 'full speed,'" he wrote. "We can only do that if
we are true partners."

Hood agreed to give it a year. But he couldn't stand to wait. By
New Year's Day of 1999, Hood was ranting again about foot-drag-
ging. He proposed taking an unpaid leave of absence. He considered
resigning.

A thoughtful compromise emerged. What if Roger Perlmutter
could be recruited to run the institute's day-to-day operations,
allowing Hood to serve as chairman? The idea had some appeal
to Hood, although he imagined it more as a partnership in which
he was the boss. He had a close relationship with Perlmutter and
respected his scientific mind. But Perlmutter already had a high-
powered, high-paid job as an executive vice president at Merck, the
pharmaceutical giant. Even so, Ramsey said, they had at least one
"very good" conversation. "Both Lee and Paul saw me as someone
who could be a peacemaker," Perlmutter said.

The regents gave their formal blessing to the dean in April 1999
to work on creating an institute. With everyone properly looped in,
the public was told of the latest world-changing vision. The *Seattle
Post-Intelligencer* called it "A Big Splash in the Gene Pool." The institute would be part of UW but located off campus. The annual budget
of $100 million would come from government, industry, and founda-
tions. "This represents the future of biomedical research. Ensuring
the success of the institute is, and will remain, among the School of
Medicine's top priorities well into the twenty-first century," Ramsey
told reporters.

Perlmutter declined to join, saying he felt "much satisfaction" at
Merck. But he expressed interest in Merck sponsoring the institute.

Ramsey and Huntsman thought the institute could still proceed. They told Hood he could be the "founding chairman":

> We strongly believe that this extremely important
> role as Founding Chairman of the Board is the
> best administrative role for you to have with the
> Institute … given your travel schedule and your time
> commitments to Molecular Biotechnology and your
> personal research. … We will ask Bob Franza to …
> assume the role of Acting Director of the Institute. It
> will be necessary to initiate a national search for the
> permanent Director.

Hood felt blindsided. He didn't want to be a figurehead. Fearful, Hood saw himself being marginalized at his own institute. He scrambled, trying to arrange a meeting on short notice with billionaire Paul Allen, the Microsoft cofounder. That irritated UW officials.

Even Huntsman, a patient man, started "sermonizing," in his own words, in an e-mail he sent to Hood that July:

> My sense is that we negotiated a pretty clear
> understanding of how the Institute will be structured
> last year. … Since then, a number of people have
> worked very hard to advance the governance, facility,
> IP (intellectual property), and business plans for the
> Institute, and the result has been excellent progress.
> During that period it seems that you have been
> somewhat disconnected and have put little effort
> into the Institute. …
>
> Paul [Ramsey] and I have tried to be clear that we
> need and want your vision and leadership of the

Institute while also needing assurance that the
operations and vulnerabilities associated with it will
be managed carefully.

You have somehow thought that we are trying to
take the Institute away from you. That is certainly
not the case. ... There needs to be a title that
acknowledges your leadership. We had thought
"Chairman" did that; if not, let's find a more
appropriate title.

But, the title is just a small part of the problem.
I think the key to moving forward is to establish
trust. You have articulated a concern about this.
... What I'm not sure you understand is that
the same question is in our minds. ... There are
several examples, the most recent of which is your
statement to me on Wednesday that you sought an
appointment with Paul Allen and will soon seek one
with Bill Gates. You know full well that that is out of
bounds behavior. The University must coordinate its
approaches to major donors. ... It's foolish behavior.

Hood replied that this was "unfair," because he gave notice of
his outreach to Allen a year earlier. But by this point, misunder-
standings were accumulating. People were talking past each other.
Exasperated, Hood agreed to one more meeting with Huntsman and
Ramsey in August. "Perhaps a miracle will happen," Hood wrote in
an e-mail.

Never much for religion, Hood started laying the groundwork for
a totally independent institute, separate from UW. Hood would be
president, accountable only to the institute's board, which he would

select. He connected with an experienced administrator named Louis Coffman for help. In an e-mail on September 20, 1999, Coffman told Hood he was ready to work on it "monomaniacally." By mid-October, Coffman found lab space and a law firm to draft bylaws. Coffman reserved an Internet domain name for the new independent institute. "No bureaucracy here!" Coffman wrote to Hood.

That same day—October 14, 1999—Hood sent a letter to President McCormick that described "failed efforts" to create an institute within UW.

> I am ... concerned about the future of the University of Washington and its aspirations for excellence and academic leadership. I use my failed efforts with the University only as an example to illuminate what I believe to be a fundamental challenge. ...

> Six universities, three public and three private, are developing research institutes to take advantage of opportunities for biology in the post-genomic era. Five universities have committed $60–$200 million to their institutes; these same five universities have made space available on campus ... and they have made significant commitments for faculty staff, and/ or administrative positions.

> Although we started discussions with the School of Medicine two and one-half years ago, the University of Washington did not make any dollar commitments. ... It made no campus space commitments, nor did it commit a single faculty, staff or administrative position to the Institute. All of these were requested.

Ramsey's response made it clear that his concerns ran deep. "In the 1960s, 1970s, and 1980s, many medical schools established a variety of private independent entities," Ramsey wrote. "These decisions led to a balkanization of the activities, and many of these medical schools are now in serious financial difficulty. In some cases, the financial problems are so serious that they are threatening the financial integrity of the parent university." Ramsey saw conflicts of interest that could undermine the mission of the university. "There is a problem of faculty members simply giving too much time to outside enterprises, diminishing the quality of their University-based research and teaching activity," Ramsey wrote.

If Hood was going to get access to the assets of the university—faculty, students, and shared resources—then he would have to be accountable to the university. "I do not believe that this proposed arrangement creates a positive model for public/private collaboration," Ramsey wrote. Ramsey ended by saying he hoped Hood and his colleagues would continue their research entirely within the university.

It was already too late.

Hood faxed a memo to Stephen Friend, the CEO and cofounder of Rosetta Inpharmatics:

> A few weeks ago, I decided to take the Institute independent from the University of Washington after months of agonizing. This decision has consequences with the University that are as of yet unclear. ... My decision to take the Institute independent is irreversible. I want to get together with you to discuss a collaboration.

Hood hadn't quit. He had one last-ditch proposal to the president and the dean: What if he split his time between an independent, nonprofit Institute for Systems Biology and his job as chairman of UW's Department of Molecular Biotechnology? He painted a rosy picture of collaboration.

It was "the dumbest idea I've ever had," Hood said years later. "There's no way I could have done both jobs. ... But I believed in it at the time. I was desperate."

Hood submitted a formal request, as required, to do the "outside professional work" of running the institute. Hood proposed transferring NIH and certain other grants, as well as intellectual property, to his institute. He inquired whether a number of university employees could be "dual employees" of the university and the institute. Others could stay at UW and serve as "volunteers" for the institute in their off-hours, Hood suggested.

This proposal was dead on arrival.

"For you to serve as a University faculty member and Chair of a University academic department, while engaging in substantially the same activities ... on behalf of a private Institute directed by you, is inconsistent with University policies and State law," Ramsey wrote.

Reading that memo, Hood felt backed into a corner. He was afraid Ramsey might accuse him of violating state ethics law. As Hood recalled years later:

I went home and really thought about it. I thought I'm really in a hard place, because if I push with this, and Paul hits me with an ethics charge, there's no doubt, within a year or two, I'd be cleared, because there wasn't anything there, but I'd never start anything again.

I decided that Paul was never really going to understand, would never be sympathetic, and that if I want to do systems biology, it's not going to be at the UW. It was the same really hard decision that I made at Caltech.

The smart and safe thing at Caltech would have been to forget about the genome project and forget about all this technology development and become a more ordinary biologist and fit in with the rest of the milieu. It's a great place, great students, has lots of money. Caltech is terrific. But I wanted to do something that didn't fit into Caltech. And here I wanted to do something that didn't fit into the UW. It was an enormously scary decision. The move from Caltech was hard because of the quality of Caltech. The move from the UW was hard because I was on my own. As long as you are an academic, you don't realize how much support you have behind you. You've got security.

Hood's letter of resignation was delivered Friday, December 10, 1999, to President McCormick, Provost Huntsman, and Dean Ramsey. It was effective immediately. The leader of a department with a $30 million annual budget was gone.

"My time at the University has been wonderfully productive; however, I need to devote all of my future efforts to the creation of the Institute," Hood wrote. "I look forward to many fruitful interactions between the University and the Institute." The letter was diplomatic in parts, bitter in others. "The University of Washington has not supported the Department very effectively," Hood said, citing space constraints that drove away faculty.

Hood suggested they appoint Barbara Trask as acting chair to replace him, transfer his grants to the institute and let him keep a

UW affiliation, follow through on the K-2 expansion, and keep supporting community outreach. "The Department is very young, and hence, very vulnerable," Hood wrote.

News reports picked up on the breakup. The university "turned out to be a bit too constrained for the envelope-pushing, entrepreneurial scientist," the *Seattle Post-Intelligencer* reported in a front-page story. All players interviewed tried to put out positive spin. Olson was gracious, saying Hood "benefitted the university greatly." President McCormick said Hood did a "brilliant job" recruiting talent and raising the university's profile in the fast-moving, competitive world of molecular biology and biotech.

Hood said he was ready to be his own boss. He told reporters he was getting started with an anonymous $5 million donation. He didn't tell them the donation was from himself. If any high-level businesspeople had been on board for funding a $100 million institute at UW, they weren't on board for this. Hood wanted to go out on a high note. He made sure to tell reporters that his department's grant support climbed from $3.5 million in the first year to $32 million in his last.

Valerie Logan had told her husband he shouldn't quit in a huff. "I pleaded with him to be a little more gracious and slow and give people notice, and he said, 'No, I'm ready to go,'" Logan said later. Hood had invited her to attend some of the tense meetings with the dean and provost. She never said a word at those meetings, but in postmortems at home, she helped Hood interpret others' behavior and consider next steps. She knew better than anyone that once Hood had set his mind to something, he would become maniacally focused and determined. "One of Lee's greatest strengths is that he doesn't recognize defeat, even when it's staring him straight in the face," Valerie once told a friend.

Hood's son, Eran, visited his parents that Christmas and recalls saying to his father: "Oh my God, what if this doesn't work? Is this a good idea?" Even in private moments with family, Hood was never prone to self-doubt. Still, this was undeniably a moment of great uncertainty. "He was definitely a little anxious," Eran said.

Looking back, there were multiple reasons such an institute could never work within the university. There was cultural resistance to breaking down walls between academic departments, which all have their own hierarchical power structures, said Alan Aderem, who joined Hood to cofound the Institute for Systems Biology. There was no way a UW-based institute could interact with companies and develop technologies in the free-form way Hood envisioned. The university was "clueless" in how to manage intellectual property, Aderem said. With its traditional, cheap graduate-student labor model, the university also couldn't afford to compete with Microsoft, across town, on salaries for technology people. And relations became toxic between Hood and Ramsey. "It was personal," Aderem said. "Lee's vision is all that matters to him. Ramsey was in the way." Years later, Hood noted that the previous dean, Phil Fialkow, was a scientist. Hood could be merciless in his put-downs of Ramsey as a mere clinician who didn't "understand" science.

Viewing the battle from afar, Bill Gates, years later, said he saw the merits in Hood's vision, but also saw why he clashed with the administration. "One thing about Lee is that he can be demanding. ... He's a wonderful guy, but a very demanding guy," Gates said. "He was ahead of systems biology. ... These things are never black and white."

At the dawn of a new millennium, at a time in life when many men of accomplishment are ready to rest on their laurels, Hood took the biggest gamble of his life.

If the twenty-first century was indeed going to be "the biology century," he would have to hustle like never before to play a central role. No longer could he count on institutional support. Bill Gates wasn't writing any big checks. Hood would have to spend his own money, risk his reputation. If the Institute for Systems Biology were to fail, for either financial or scientific reasons, it would be an ignominious end to an incredible career.

For Hood, running a start-up was uncharted territory. He'd advised biotech start-ups, but always as an outside observer, not the company builder. This time he was running the show. He knew the odds were against him. Once, in a meeting with Monsanto, he muttered that bootstrapping an institute on his own dime was nutty. "Those never work. You can't get anything done," he said. Yet there he was, doing exactly that. If he failed, he'd have no one to blame but himself. It was terrifying. And, Hood said, "it was enormously exhilarating."

One former postdoc sent Hood a card; amused, Hood replied with a handwritten note of thanks. The card was a picture of a frog jumping into a frying pan.

Chapter 15

Start-up Days

The first morning of the twenty-first century in Seattle was an ordinary one: chilly and drizzly. Hood's mind was racing. What to do first?

He wasn't the type to be overwhelmed. At the age of sixty-one, he had good reason to believe he could barrel through everything that needed to be done. He could figure out what he didn't know on the fly.

The timing for starting his own institute was good. The Y2K computer bug didn't create an apocalypse, as feared. Technology investors were swimming in paper wealth from the dot-com boom. Microsoft had minted thousands of millionaires, before the rise of Apple and Google. The media told countless stories of teenagers getting rich from day-trading in tech stocks. Surely, Hood thought, he could raise money in such an environment.

The political wind was blowing in the right direction, too. The National Institutes of Health budget doubled from 1997 to 2002. The Human Genome Project, driven to compete with Craig Venter's Celera, was galloping along. Always thinking a step ahead, Hood knew that the government would need to adjust research-funding priorities for the post-genome age. "Systems biology" was still a fuzzy buzzword to many, but Hood sought to define it as the logical next step in biology.

For fifty years, biology had been on a quest to drill ever deeper into the mechanics of genes, proteins, and cells. This fundamental drive was known as the molecular biology revolution. Its origins traced back to Watson and Crick's seminal 1953 discovery of the DNA double helix structure. Systems biology, in a lofty sense, represented a new chapter in the history of biology. This was about taking what had been learned about the individual trees and seeing how those trees fit into the broader ecology of an entire forest. This was about studying multiple genes, multiple proteins at once, and seeing how they operated in concert. This was about both zooming in with a new focus on network connections between component parts, and zooming back out to find what patterns you might see in those networks.

Hood described the new Institute for Systems Biology in a January 2000 mission document:

> Traditionally, biology has focused on studying
> individual genes, proteins, and organisms
> in isolation. Systems biology, however, has
> demonstrated that biologically significant
> phenomena arise from complex interactions of
> numerous gene, protein, and cell elements that
> form informational networks and systems. New
> technologies have opened possibilities of recording

and analyzing cooperative activities of the elements
that form the codes of life.

Identifying biology's component parts, mapping their connections, and using computers to make predictive models of how everything might behave in concert of whole organs or whole organisms?
That was an enticing, futuristic vision. To many biologists, it was so
complex as to be science fiction, a nonstarter.

But Hood would hear none of it. He was enraptured. He also
made clear this wasn't just science for science's sake. In that same
founding strategy from January 2000, Hood promised to turn the
health-care system upside down with science that focused on pre-
venting disease, not just reacting to it. He wrote:

> The Institute will develop the tools necessary to
> advance preventive medicine—the ability to analyze
> an individual's genes, identify disease predisposition
> and use preventive techniques to block the onset
> of diseases such as cancer, heart disease, and
> autoimmune diseases.

High-level strategy was his comfort zone, but Hood's job was
now about more than that. Organizational details, never his strength,
were vital. Hood needed to demonstrate he had a legitimate nonprofit—with bylaws, a mission statement, a board of trustees—if he
was going to transfer grants from the University of Washington. This
needed to be done quickly.

Just as important were the people. Graduate students and postdocs were moving fast on their projects; they didn't want to lose
momentum and come in second to competing labs. Hood counted
on these young troops to execute on his ideas. Most were loyal.
Like star college basketball players who are drawn to play for a

great coach, the scientists were likely to follow him because they had a stronger bond with him than with the university. Even so, he couldn't dawdle. The roughly fifty people in the lab knew that it would be good for their careers to stay in Hood's orbit—but not if it meant they'd struggle to do science. Hood also didn't want to provide an opening for someone else to seize the first-mover advantage in systems biology.

Speed was of the essence. With no committed investors, Hood laid down $5 million of his own money. That was the anonymous donation he told reporters about, as he was perhaps embarrassed to say he hadn't raised money from his wealthy technology friends.

Hood could afford it. He was pulling in hundreds of thousands of dollars a year in royalties from automated DNA sequencers. He'd held all of his Amgen shares throughout the company's meteoric rise. Shares of Amgen rose 5,788 percent in the 1990s, making it one of the ten best stocks of the decade. But now, Hood thought most of the science at Amgen was mediocre. George Rathmann, the leader he admired, was long gone. It struck him as a great time to cash out.

Just at the moment he needed to stir goodwill among potential donors, Hood took a hit to his reputation. The *Los Angeles Times* ran its exposé with Henry Huang's claim of co-inventorship of the automated DNA sequencer. And had Hood secretly used federal funds to develop the automated DNA sequencer at Caltech in the 1980s? If so, he and Caltech might owe royalties. On February 16, 2000— during the earliest, most vulnerable days of the fledgling institute— the *Times* reported on the investigation by the inspector general of the US Department of Health and Human Services. Hood had tried to secure federal funding for the DNA sequencer, but when his pleas fell on deaf ears, he turned to private sources. To be accused of cheating the feds out of their share of royalties—after the government repeatedly shot down his idea—was galling. Gossip was already circulating in wealthy tech circles about Hood's loose-cannon ways and what really happened behind closed doors at the University of Washington.

Hood was livid. "What [federal investigators] are doing is wrong," Hood told *Science* magazine. "It's a witch-hunt."

Hood, with his own lax attitude toward accounting, had done himself no favors. Federal funds did arrive belatedly; had they played a critical role in finishing what private funds had started? Questions about the timing of improvements to the instrument were legitimate. Perhaps out of courtesy, or an effort to curry favor with grant makers, Hood even publicly gave some credit to the National Science Foundation for contributing to the DNA sequencer at his 1986 press conference at Caltech.

Through an affidavit from an NSF official, Hood clarified that the agency started supporting the instrument only after an early working prototype was built and the key intellectual property had been developed.

The *Los Angeles Times* wasn't buying this. It kept the heat on with an investigative follow-up in May, headlined "Scientists Insist They Invented Technique without Federal Funding, but Paper Trail Suggests They Relied on Millions in Grants. Inquiry Will Determine if There Is Need for Payback."

Hood sat through hours of depositions, in addition to meetings with Caltech attorneys, to resolve the matter. Hood's personal attorney, Mauricio Flores, said his client remained an ideal witness—consistent, articulate, unflappable. Federal officials dropped the investigation a couple years later. Hood said the newspaper owed him an apology. It never followed up to report that the charges were dropped.

The Institute for Systems Biology, despite the legal distraction, charged ahead at full speed.

Hood wooed two cofounders who couldn't be more opposite: Alan Aderem and Ruedi Aebersold. Both were entrepreneurial spirits from the younger generation, and among his favorite professors at the University of Washington. Aderem and Aebersold had been part of talks about the institute-within-the-university idea. They knew Hood's vision for systems biology, how their science fit into it, and why such an institute couldn't work inside the university. They were ready to join him as cofounders.

Aderem, a stocky man with long curly brown hair, was a mercurial and passionate character. He had gone to prison as a younger man to protest Apartheid-era policies in his native South Africa. A highly regarded immunologist, he had a first-rate pedigree from Rockefeller University. He had been recruited by Roger Perlmutter to the University of Washington a few years earlier, joining in 1996. He came partly because of the Seattle lifestyle and partly because he was exhilarated by a long talk with Hood. They found they had a shared vision for systems biology after the Human Genome Project.

Aebersold was a different personality altogether. Level-headed. Stoic. A stabilizing personality in the go-go early days of the institute. A native of Switzerland, he grew up scientifically as a cell biologist in the more reserved European tradition. He provided emotional ballast for the overexuberant Hood and the moody Aderem. If the founding team drew up a list of ten compelling projects and had the resources to do only five, Aebersold would rigorously walk through the priorities in disciplined, stepwise order, even if Hood preferred to ignore all budgetary constraints.

"The guy who was steady on the ground was Ruedi," Aderem said. "He is a rock."

Aebersold also was a world leader in proteomics—the study of the structure, function, and interactions of proteins at large scale. Aebersold had done a postdoctoral fellowship under Hood during the 1980s, proved himself as an independent researcher at the University of British Columbia, and later joined the Molecular

Biotechnology faculty at UW. He brought essential expertise. After all, once you had the codes for all of those genes, biologists naturally wanted to know how the codes translated into the proteins that do most of the work in cells. Proteins were hard to study at large scale. Their function also couldn't be deduced automatically from looking at their underlying sequence. It depended in large part on how they folded into 3-D shapes. Given all the variables, scientists tended to look at proteins one at a time. Aebersold's work promised to make it possible to evaluate many different proteins at once, at automated high throughput, the way genomic analysis had been transformed.

Louis Coffman, the man with organizational skills the scientists lacked, was another crucial member of the founding team. An accountant by training, Coffman gained valuable experience as chief operating officer of the Burnham Institute for Medical Research, a nonprofit research center in San Diego. He had left that job in 1998 to do some independent consulting. He was intrigued when he got the call from Hood.

Coffman, an affable character with a bushy mustache, happily took on the role of "chief cook and bottle washer." He found cramped lab and office space on Seattle's Roosevelt Way, walking distance from the university. In a weird twist, the university was the institute's landlord. Coffman transferred grants, set up payroll systems, and bought furniture and lab supplies. He coordinated the moving trucks, set up proper workplace safety procedures, and made sure all policies and paperwork necessary for federal grant recipients were in place. Questions at UW about how to handle Hood's grant transfers, and who owned certain intellectual property—Hood or the university?—caused some delays. With his family in San Diego, Coffman worked around the clock. He practically moved in with Hood, staying at the guest cottage next to Hood's dream house on Lake Washington.

Coffman remembers the adrenaline-soaked start-up days with fondness. There was a "Three Musketeers" attitude: all for one and one for all. "I absolutely loved it," Coffman said. Once, when Coffman discovered Aderem needed a new cell phone to do institute business, he dashed to a store, bought one, and handed it to Aderem within an hour.

Hood succeeded in getting his grants transferred. But the university declined to enter into a master affiliation agreement with the institute. Such an agreement would spell out terms for how faculty members could go back and forth from the university to the institute, how salaries would be paid, how the intellectual property they created would be handled, and how grad students were managed. Without a master agreement, Hood was forced to hash out smaller, more limited governance deals with individual UW departments: computer science, microbiology, biochemistry, immunology, bioengineering. Even years later, UW maintained a frosty relationship with the institute. Some UW professors discouraged students from doing postdoctoral work there.

Looking back, Aebersold said, the fallout with UW may have been a blessing in disguise. The founders knew they needed to make the institute work on their own.

"I'm convinced that if we had a fallback position, and said to ourselves, 'If it gets hard over there [at ISB], we can go back to 100 percent at the UW,' then the institute probably wouldn't have made it," Aebersold said. "It forced everyone to go for it. There was no going back."

Lacking a tight affiliation with a major national research center meant the little research institute would essentially be on its own when recruiting. New faculty would have to draw their whole salaries from one institute, and maybe teach a class once in a while at the big-name research university. It was fundamentally a different employment proposition from having tenure, "the closest thing you could get to lifetime employment," Coffman said. The institute offered no such security.

Hood wanted it that way. He thought tenure enabled people to rest on their laurels. Salaries at the institute were set high to compete for top talent, but nobody could get too comfortable. Instead, faculty would be on rolling, renewable three-year appointments. They'd be responsible for securing their own grants. Hood retained the authority to hire and fire people at will, and the flexibility to pay above-market salaries if needed. These early decisions made the institute more like an entrepreneurial biotech company than a university. As Hood put it:

> I wanted to really push the idea of being an
> entrepreneur, in the independent spirit. I just
> never agreed with tenure. ... Tenure does, for many
> people, breed complacency. I didn't want any people
> that tenure could be a crutch for. One way to do that
> was to do away with tenure. It weeded out people,
> absolutely.

It also weeded out the ability to recruit some truly world-class talent. People who already had tenured positions at top institutions were unlikely to give that up, especially for what amounted to a start-up with no major financial backing in a gray, rainy place far away from many scientific collaborators. This wasn't going to be like the Whitehead Institute in Cambridge, Massachusetts, where faculty would have easy access to students and prestigious collaborators at MIT, while also enjoying the benefits of an independently funded research lab.

These were obstacles, but not insurmountable ones. Hood had recruited many excellent scientists to the University of Washington, and some chose to follow him. "I was way too engaged in what I was doing," said Trey Ideker. "I knew I wanted him to be my adviser. There was never any question about switching advisers." He moved to the institute for one year, where he finished his PhD in molecular biotechnology. Through an affiliation agreement, his diploma says it is from the University of Washington. Even if the powers that be at the university disliked Hood, they weren't going to punish his students if they followed through and did the necessary work to get their degrees.

UW President McCormick offered some diplomatic comments for a story by the *Chronicle of Higher Education*, a widely read publication in academic administrative circles. "The institute will have very close ties to the university and draw on the intellectual strengths of the university, many of which Lee Hood put in place," McCormick said.

Money was never far from Hood's mind. He himself took zero salary the first two years and only $88,000 in the third year—less than half what cofounders Aderem and Aebersold were being paid. Repeatedly, he had stated a goal of building an endowment of $100 million or $150 million. On his most optimistic days, he'd blurt out that his goal was $200 million. That kind of financial cushion, invested wisely, could throw off $10 million a year in earnings alone. With that kind of discretionary cash flow, Hood could readily advance his most farsighted ideas. Without an endowment, the institute would have to live paycheck to paycheck. It would need to focus on winning competitive grants and contracts, which almost always depend on hitting short-term, incremental goals set by corporations and government funding agencies.

Hood had a long donor list, but he was fixated on just two names: Bill Gates and Paul Allen. The billionaire cofounders of Microsoft were both intellectually curious. They knew Hood. And they had both made money in one of Hood's start-ups, Darwin Molecular. Gates was the most likely benefactor, because the two men had developed something of a friendship. Then again, Allen was the one with the appetite for the most futuristic and risky initiatives.

Hood's administrative assistant, Tawny Biddulph Burns—whom he lured from the university—set up the meetings. She remembered Hood as a humane and conscientious boss toward low-level staffers. "He offered me a raise. He made sure my benefits package was equal, if not more than, what I had at the university," said Biddulph, who worked thirteen years as Hood's assistant. "Lee would tell me what he needed done, and I'd get it done. I liked the responsibility, the trust."

It was still early 2000 when Hood put together his best sales pitch for Gates. But suddenly the timing was all wrong.

Gates was battling an antitrust case. The federal government was threatening to split Microsoft into two companies for what was perceived as predatory, monopolistic behavior. Even though Gates had been dismissive of Hood's institute idea at first, he'd agreed to meet with Hood to hear more about it.

The three cofounders, along with Coffman and Hood's wife, piled into Hood's old Toyota Camry. Everyone brimmed with anticipation. Aderem and Aebersold had never met Gates. Hood assured his fellow scientists that the tech mogul was a friend and surely would become a financial supporter. Gates was no biology dilettante. Hood told his cofounders to be ready for piercing, shrewd questions. Logan and Coffman waited in the car while the three scientists walked into Gates's office at Microsoft Building 8. The meeting lasted a little more than an hour.

Things got off to a good start, Aderem recalls. The scientists showed an attractive PowerPoint presentation with some data from Hood's work on prostate cancer. The research held promise to someday untangle the complex biological networks that allow tumors to grow and spread. At one point, Gates wondered out loud whether this kind of more holistic "systems" approach to disease might have helped his mother, Mary Maxwell Gates, who died of breast cancer in 1994.

After an hour of questions and answers, it was time to talk business. Gates asked how much money the team wanted. Aderem and Aebersold have differing recollections of what Hood said next, but he blurted out either $100 million or $200 million. Both men recall Gates being startled.

"Lee, that's a lot of money," Gates said.

After an awkward pause, Gates had another question. "Have you talked with Paul [Allen]?"

No, they hadn't. The men exchanged pleasantries on the way out, after what had been a stimulating back-and-forth. No promises were made. No checks were written.

Nonetheless, Hood brimmed with optimism. Gates hadn't said no. One more meeting ought to do the trick, he told his compadres on the ride home. The cofounders weren't so sure. They saw a brush-off and a strategic misstep. "Lee didn't want to be nickeled and dimed," Aebersold said. "He wanted one pot of $100 million, instead of twenty to thirty pots of $1 million to $2 million each. In retrospect, that was a mistake."

By mid-2000, Hood was banging out a steady drumbeat of press releases to show that the institute was real, not just his imagination. Perlmutter at Merck agreed to sponsor $5 million of research over five years. That July, Hood announced that Aderem, the immunologist, and Aebersold, the proteomics scientist, had joined as cofounders. Perlmutter joined the board, as did George Rathmann, the former Amgen CEO, both adding gravitas. The original scientific advisory board included stars like Sydney Brenner, Lee Hartwell, and Lee Silver. Two other faculty members had arrived—astrophysicist George Lake and cell-sorting technologist Ger van den Engh, from Hood's department at UW. The eclectic talents underscored Hood's message: cross-disciplinary thinking was required for the new biology.

Morale ran high. Unleashed, Hood was at his brilliant best, firing off what seemed like a hundred ideas a day. Anything and everything seemed possible, said John Aitchison, a young scientist recruited from the University of Alberta. "I remember one early meeting was fantastic at Lee's house on Lake Washington," Aitchison said. "We basically said if you could do anything you wanted, what would you do?" Hood and his troops felt in a position to make history, to set the agenda in the aftermath of the Human Genome Project.

Aderem, an intellectual sparring partner with Hood, had seen his friend get worn down by the old university battles. Venturing out on his own stirred the man's creative juices. Not only that, Aderem believed Hood's was a personality that thrived on danger. He was like the high-wire walker who focuses entirely on the task at hand, because the alternative is death. Aderem observed:

> If you put Lee in a corner, he pushes back. He becomes less creative. He needs that absolute chaos around him to be creative. He needs to break every fricking rule in the book to be able to harness that creativity.

Cytoscape was one of their first projects. The computer program helped biologists better visualize biological networks—which were hopelessly byzantine not only to the uninitiated but to many scientists as well. Hood and Ideker coauthored a manifesto of systems biology thinking. "A New Approach to Decoding Life: Systems Biology" was published in 2001 in the *Annual Review of Genomics and Human Genetics*. That same year, they published a paper in *Science* on how complex systems respond to perturbations, at least in a relatively simple organism: yeast. Getting a paper in a top-notch journal like *Science* was always cause for celebration, but it was especially important for a new institute striving for peer validation.

The culture mirrored that of the tech start-ups of the day. Work hard. Play hard. Don't take yourselves too seriously. A poster was drawn up depicting the founders as "The Three Amigos," complete with sombreros, bolo ties, and pistols at their side—like the characters in the 1986 movie comedy starring Steve Martin, Chevy Chase, and Martin Short. The classic structure in science, of independent investigators pursuing independent ideas, was out. The founders talked about team projects that would foster a collaborative culture. They decided they'd combine efforts to write a textbook. The idea was instantly embraced. Ideally, it would help them sharpen up their thinking about systems biology, while staking the institute's claim as a pioneer in the field.

The cramped, garage-like space also resembled that of the start-ups of the day. Aitchison set up his desk underneath a stairwell, next to a boiler—at least for a few days, until he got sweated out. Aderem shared his office with four other people. It was a "rabbit warren," Hood said. Hood brought over almost fifty people from his old lab, and the staff swelled to more than eighty when Aderem and Aebersold's groups arrived. Before long, people had to wait in line to go to the bathroom. But being crammed together was inspiring, too.

Scientists from those days fondly remember having stimulating conversations at all hours of day and night. Gates would occasionally stop by, security in tow, to talk about science. "I remember once I was telling him about the 3-D organization of chromatin in the nucleus," Aitchison said. "He'd just come and learn biology."

Gates didn't open his checkbook, though.

Across the street, the University of Washington was determined to maintain its world-class position. Replacing Hood with someone of similar scientific stature was critical. A national search for a new chair of genome sciences netted Robert Waterston of Washington University in St. Louis—a quiet but important player in the Human Genome Project. Lifers like Paul Ramsey, Lee Huntsman, and

Maynard Olson were committed to continuing great work. They scrapped the name "Department of Molecular Biotechnology," a brand virtually synonymous with Hood. Olson, Phil Green, Stan Fields, and Debbie Nickerson were brought together with more classic geneticists, like Mary-Claire King, in a new "Department of Genome Sciences." By April 2003, Ramsey had some big news.

Bill Gates was giving $70 million to the university for a state-of-the-art building to house genome sciences and bioengineering. It was Gates's biggest gift to the university, dwarfing his 1991 donation of $12 million to lure Hood.

The donation could have been interpreted a couple ways. It was a testament to the people and momentum Hood brought to the university. It could also have been seen as a slap in the face to Hood. When I reported the story for the *Seattle Times*, Gates never mentioned the former chairman of molecular biotechnology by name. He focused on the community instead:

> This will cement the leadership role the university is in, and holds great promise for pushing forward the contribution that genomics work will make to all kinds of diseases. ... The way it works is a few places gather a lot of the great scientists, and that's where they get the grants and the equipment. You end up with three or four places that really do the best work. Almost anybody would put Seattle on the list in genomics as one of the three or four places where there's a lot of work going on. It's a leader, and I believe this community will stay a leader.

Five years passed between Hood's first request and Gates's first grant, in October 2005, to the Institute for Systems Biology. The donation was a relatively small $10 million "challenge" grant. "Lee has built an impressive team," Gates said in a press release. "With this grant, we are supporting an innovative Northwest-based organization that offers unique potential to combine world-class medical research and technology and change the way we think about predicting and preventing disease."

It appeared to be a strong statement of support. But a challenge grant required matching funds. Gates would provide the full $10 million only if the institute first rounded up $100 million from other sources. This was a stingy matching grant: one dollar from the Gates Foundation for every ten dollars raised from other sources. It took several more years—and an unorthodox decision from the tiny government of Luxembourg—to put Hood in position to collect the full amount from Gates.

A decade later, Gates reflected on the two men's divergent interests in biology. The Bill & Melinda Gates Foundation marshaled its resources against scourges overlooked by the pharmaceutical industry—particularly HIV, malaria, tuberculosis, and diarrheal diseases. Hood's work in systems biology was more relevant to common ailments in wealthy countries. "We love Lee's work, but it's not in the foundation's sweet spot," Gates said in an October 2014 e-mail. "Anything we do with Lee is me just funding a guy I know and really admire who does great basic research."

A $200 million endowment was not going to materialize anytime soon. That much was clear. The institute ran in the red, struggling at times in the second year to make payroll. By 2001, the institute took out a $9.25 million commercial loan from the Bank of America to cover salaries and other operating expenses. The loan had to be paid back in three years. Hood guaranteed the loan by offering some of his remaining Amgen stock as collateral, and the loan-repayment deadline was later extended to 2009.

Coffman said the Bank of America loan made him nervous. If Amgen stock dropped, the bank could have forced Hood to sell shares to pay back the loan. It could have left Hood "horribly exposed," Coffman said. "He said at one point, 'Louie, just help me and Valerie so we don't end up on the street,'" Coffman said. Hood was only half joking. Coffman crafted a plan that allowed Hood to donate Amgen shares to the institute in such a way that the founder wouldn't be personally liable if the institute went bankrupt.

The *New York Times*, intrigued by the new venture of a "biotech superstar," detected considerable skepticism in the science and business community. The *Times* quoted Maynard Olson: "He starts things. That's what he does. There is a trail of acrimony in most of these cases." Another leading scientist acknowledged Hood's vision in a backhanded way. "He's an opportunist, a scientific opportunist, in that he knows what to pick up, and I think that's important," Sydney Brenner told the *Times*. Gates declined to comment.

Hood began traveling the world again. This time, he was searching for partners and financial backers. With characteristic intensity, Hood would take red-eye flights from Seattle to the East Coast, land at 6:00 a.m. or 7:00 a.m., and go straight to the meeting without checking into a hotel or taking a shower. Once, Aderem recalls sitting at the back of a plane with Hood, in narrow coach seats, when Hood fell asleep on his shoulder. Drifting off, Hood clutched a small cup of coffee that was dribbling on his sleeve and pants. These were the same clothes he was planning to wear to an important meeting. There might be time to touch up the spots in the airport bathroom, but that would be it. "I looked at him and said to myself, 'This is what success looks like,'" Aderem said. "That's who Lee is. He doesn't give a shit about that kind of stuff. It's a lack of pretentiousness at the deepest possible level you can imagine."

Hood continued to personally finance the institute. By 2002, as the employee count grew to more than 170, the organization moved out of its rabbit warren on Roosevelt Way to a more modern space, twice the size. The building was beautiful—designed for collaboration, with open spaces, natural light, and a brightly colored paint scheme—with sweeping views of Lake Union, the downtown skyline, and the Space Needle. Hood and the founders let the young people take the views. Hood's corner office window looked out at a local bakery and the municipal dump.

Part of the money for the facility came from a lucky break. Rosetta Inpharmatics, one of the companies Hood cofounded, went public in 2000. A year later, Merck acquired it for $620 million. Hood may have been asked to leave the scientific advisory board after buying a competitor's instrument over Christmas break, but he still had founder's shares. The deal put about $6 million into Hood's pocket. All of it went into the institute. Hood was also getting royalty checks from the automated DNA sequencers sold by Applied Biosystems. And he started winning various invention prizes for the DNA sequencer, which came with sizable cash awards: the Kyoto Prize, the Lemelson-MIT Prize, the Heinz Award, the Kistler Prize, the Russ Prize. Together, these provided more than $1.37 million— a bigger haul than the Nobel Prize. Every time Hood won another award, he plowed the proceeds back into the institute. Hood's total investment in the institute was more than $25 million in the first five years.

Hood's failure to raise the endowment puzzled many scientists at the institute. He had such a strong track record as a fund-raiser. He had a golden Rolodex. He could still give spellbinding talks. But several factors worked against him.

The dot-com stock boom went bust. The genomics stock bubble also popped, after everyone realized that Craig Venter's Celera Genomics couldn't sell subscriptions to its proprietary genomic database if pharma companies could simply turn to a free, public alternative from the Human Genome Project. The market collapse began just a few months after the institute got started, and many paper millionaires suddenly felt poorer. Many hyperbolic things had been said about the Human Genome Project revolutionizing health care, and donors saw that wasn't going to happen as quickly as promised. The growth of the NIH budget was also a two-edged sword. The larger pool of grants made it easier for the institute to win the support it needed. But some prospective donors wondered whether their support was essential or just icing on the cake. And when federal budgets plateaued in the mid-2000s, the institute had to rein in its ambitions. Its budget flattened out at roughly $20 million a year in grant support.

Hood was able to arrange meetings with prominent donors and outline his vision, but, in Coffman's view, he pitched to them more like investors than donors. An investor wants to know if he writes a check, what's in it for him. Hood's advisers pressed him to make a traditional nonprofit pitch, asking people to support the institute's long-range, public-service-oriented mission. Hood didn't quite do that. He described the institute almost like a company and he was "selling stuff," Coffman said. But he wasn't selling something that would make people wealthy. Since the Institute for Systems Biology was a nonprofit, no one was getting an equity stake in some hot for-profit start-up. It wasn't a compelling pitch.

Strategic partnerships weren't panning out, either. The process was time-consuming and, at times, made Hood appear painfully naïve. Looking back, Hood said he considers much of the effort a waste of time. He spent more than five years seeking partnerships with government officials in Israel, Ireland, Alberta, South Korea, and elsewhere. In Israel, for example, he spent two years developing a partnership, traveling to all seven major research centers in the country to sell it in person. The idea was for an Israeli fund-raiser to pull together $100 million for a systems biology center at one of the Israeli centers of excellence, and an equal sum for the Institute for Systems Biology, so they could collaborate on big genomic and proteomic projects, hopefully spin out companies, and enjoy shared prosperity. After two years of planning, the fund-raiser came to Hood and said he could corral the $100 million for an Israeli institute, but no money could go to the institute in Seattle. "Thanks for telling me now, after I've spent two years working on this thing," Hood told the financier. He walked away, frustrated.

Ireland was interested in a smaller two-way partnership, thanks to a former National Science Foundation official who took a job in the Irish government, and who was an acquaintance of Hood's. But talks petered out after a few months when one of the local research centers decided it could do a systems biology program on its own. Alberta was a different story, awash in oil money from the Canadian tar sands. Local officials were looking to invest in a clean industry of the future. Exciting meetings led to more exciting meetings, but ultimately, no one could ever make a decision, Hood said. Korea was all set to go until the incumbent government Hood had been negotiating with lost an election. Each negotiation typically took nine to twelve months—time away from science, colleagues, and strategic decisions. "Lee, you're wasting your time," advisory board member Perlmutter said on more than one occasion. "Spend your time in Seattle." Years went by before Hood acknowledged that the partnership efforts had been misguided.

Hood created his own platform for spinning out companies: a start-up incubator called Accelerator Corporation. He persuaded a friendly group of venture capitalists—including his former postdoc Michael Steinmetz, now at MPM Capital—to establish the for-profit entity in 2003. The idea was that the venture firms would invest in raw ideas that had broad potential business applications but were too risky for mature companies to take on after the genomics stock bust. Hood could spin some of his best ideas out into Accelerator, and the Institute for Systems Biology could keep founder's shares. If one of the fledgling companies turned out to be another Amgen, then those shares might be used to create an endowment.

Investors put $15 million into the start-up incubator at first. Several of the little companies reached their preliminary scientific goals and went on to raise sizable sums of venture capital. But none delivered the kind of breakout success Hood had enjoyed earlier in his career through his associations with Amgen, Applied Biosystems, Darwin, and Rosetta. At a minimum, it kept Hood's profile high in the business world, showing he still had a keen eye for good ideas, even if they didn't all pan out in the end.

Money wasn't materializing, but the institute's reputation and influence grew. They got papers published in scientific journals, slowly building a serious body of scientific work. They popularized the term "systems biology," influencing fellow biologists to make it mainstream. By paying better-than-average wages and benefits, the institute made a number of "best places to work" listings in scientific magazines, which helped with recruiting. More than 150 other institutes or centers for systems biology were set up by universities around the world—the sincerest form of flattery. NIH money continued to flow to anything with the term "systems biology" in it.

The Institute for Systems Biology was ranked No. 1 among all US research institutions and No. 3 worldwide in terms of the impact of its scientific publications, according to an analysis of scientific papers published from 2003 to 2007 by a Spain-based consulting group. Impact was measured by total research output, how much the group collaborated, the influence of its publications, such as how often papers were cited by other scientists, and other factors.

Tireless, Hood continued to spread the scientific word. He captivated the science media and scientific audiences with his optimistic visions for the future of biology and medicine. When he won the Kyoto Prize in 2002, local high school and college students lined the streets next to his hotel to watch him travel by limousine to the event.

"I felt like a rock star," Hood said. Kyoto organizers asked him to give an autobiographical speech, tracing his career in science, with an eye toward setting an example for younger generations. Biotech entrepreneurs clamored to get Hood to join their boards or scientific advisory boards, even if it meant getting just a few of his hours a year. Often, they just wanted to be associated with his big name, to paste it on their website. Hood almost always said yes.

Hood hired a Clinton Administration veteran, Gretchen Sorensen, to help him with public relations: explaining systems biology and burying the past at UW. Almost every publication that called, whether top-flight or relative rag, Hood would talk with, pose for, and educate. Invariably, the coverage reflected his upbeat persona. *Fortune* magazine once featured him in a "Heroes of Manufacturing" issue, hailing him as "the man who automated biology." The *Economist*, a savvy business publication, gave him some golden publicity in a September 2005 feature titled "Medicine without Frontiers." It quoted Hood as saying that within ten years, nanotech devices would be able to analyze a thousand proteins in a droplet of blood. That droplet would contain information for a molecular "fingerprint" that would enable a predictive, preventive, personalized brand of medicine. That didn't happen. But it hardly mattered.

Americans were fed up with a dysfunctional, expensive, and reactive health-care system. The vision of new biology driving predictive, preventive, personalized medicine was enthralling. "The next goal of DNA research makes the breakthroughs of the past few years look like high school biology," *Business 2.0* trumpeted. "Some even say it's impossible—which is one reason why Lee Hood is determined to go for it."

Editors at scientific journals asked him to write articles about his career, which forced him to reflect on a life that had been spent almost entirely looking ahead. Mostly, he offered up a romanticized version. His life story was another way to engage people in the new biology. In one autobiographical article published shortly before he turned seventy, Hood wrote:

> I grew up in small towns in Montana where my
> parents always encouraged me to do well in school
> and gave me the freedom to explore many different
> dimensions of life. … I began to appreciate the
> beauty of chemistry in high school when, as a senior,
> I helped teach a sophomore biology class.
> I remember teaching the class using a 1956 Scientific
> American article on the structure of DNA. That
> article catalyzed the realization that the core of
> life was the DNA molecule with its fascinating
> chemistry of complementarity. Although I doubt
> I fully understood what that meant at the time,
> it was clear DNA was a beautiful molecule. This
> awakening pushed me toward the study of biology.

Graduate students and postdocs invariably trying to find their own career paths forward in science were eager to hear how Hood had charted his career. Hood, the consummate teacher and recruiter, was always happy to tell his personal story and to place his journey into a broader scientific and social context. In his 2003 acceptance of the Lemelson-MIT Prize, Hood said:

> Science is a most wonderful profession, where one
> may not only explore the frontiers of knowledge, but
> also create new frontiers. I have been fortunate to
> grow up scientifically at a time when biology was ripe
> for integration with new technologies. Important
> technologies in biology are all about pushing back

the frontiers in deciphering new types of biological information. ... Technologies have catalyzed major paradigm changes in twenty-first-century biology and medicine: systems biology and predictive, preventive, and personalized medicine.

Living organisms are composed of biological systems like the brain and the heart. Systems biology provides powerful new methods for studying these systems. Predictive and preventive medicine promises to extend the life-span of the average individual by ten to twenty years within the next twenty-five years, and do so by avoiding most common diseases.

It poses challenging social and ethical issues for society (e.g., genetic privacy). Scientists must bring an understanding of science to society so its citizens can respond in a rational and thoughtful manner to the opportunities for all peoples of the world.

Science education for kids remained a passion. It was a working part of the Institute for Systems Biology, not just an occasional volunteer effort. Even if Hood had to pay the bills himself. He was deeply dismayed by a rising tide of anti-intellectualism in the United States. Polls consistently showed that Americans wanted stronger education in science, technology, math, and engineering, but political support for it was weak.

Hood cringed as major public debates about scientific subjects—climate change, vaccines—devolved into ideological, political arguments that were unhinged from empirical data. One survey of more than thirteen hundred US adults by the Museum of Science and Industry in 2008 offered some typically sobering findings about the state of scientific illiteracy in the States. When asked who the science role models are for the youth of America, most respondents offered only general comments like "teachers" or "astronauts." About 6 percent said "Albert Einstein" or "Bill Gates" or "Al Gore." Stephen Hawking was the only actual living scientist to make the list, mentioned by 2 percent of respondents. Nearly half of Americans in the early twenty-first century denied the existence of evolution, and only one out of seven citizens said they personally knew a scientist.

Hood was never one to give up what he considered to be a good fight. The Institute for Systems Biology came to employ a staff of seven professional educators over the years, led by a director, Dana Riley Black, who was charged with hunting down grant support. Hood continued to testify at Seattle School Board meetings. He carved out time to have one-on-one meetings with every incoming superintendent of the Seattle Public Schools. Upon meeting one, Hood blurted out, "You're the seventh superintendent of Seattle Public Schools I've met." Black, seeking to avoid an offense, clarified that Hood meant to say that he was committed for the long haul.

There was no doubt about that. Hood knew that gains in science education might take decades to materialize. This was one endeavor in which he was patient. He thought of schools in complex systemic terms, like the human body. Slight perturbations to a single input variable (improved teacher training, for instance) might be thought of like slight perturbations to a single input variable for a human body (eating broccoli instead of cheeseburgers). Those inputs might have a subtle impact on the individual for the short-term, but they could lead to a profound improvement in the health of the larger complex system over a long period of time.

"Education is the soul of the ISB," Hood wrote in one institute report.

Valerie Logan oversaw science education work at the institute for most of a decade.

She remained Hood's most important confidant. But they also led highly independent lives. With her husband frequently on the road, Logan had her own circle of friends, and she would often go off for days on hiking or kayaking trips. By 2005, Logan had grown weary of taking care of the house on Lake Washington and cleaning up after Hood and his relentless parade of guests. One day, she told Hood she was moving downtown. He could join her if he wanted.

They sold the dream house for $4.3 million and bought a twenty-third-floor penthouse condo in Seattle's Belltown neighborhood for $3.6 million. The place offered gorgeous west-facing views of Puget Sound and the Olympic Mountains. There was room for their growing art collection, including the hulking mahogany sculpture from the Lake Washington house. Hood and his wife enjoyed walking around downtown to restaurants, theaters, the art museum. There were gritty parts of the neighborhood then—drugs, graffiti— but the overall vibrancy of the place struck them as something that, in their late sixties, kept them feeling young. They imagined retiring there, with occasional visits to their vacation home in the San Juan Islands, north of Seattle. They had some happy memories. At least until Valerie's memory began to fail, taking with it the most important stabilizing force in Hood's life.

Chapter 16

A Visionary, Not a Manager

Valerie Logan had been the one who wanted to move to downtown. Whenever the family had moved in the past, Valerie took on the logistics with gusto. Hood could keep his eye on the science, as always. "Val had been the master organizer, the master doer," Hood said. "She really enjoyed it."

But this time she was flustered. She shied away from the work. Not only was she forgetful about where things were around the house, she didn't know where to look. Hood made an appointment at a memory-loss clinic. Valerie got the grim diagnosis: Alzheimer's disease.

Hood knew that meant a long, progressive decline. Slowly but surely, his wife would lose her memory, her personality, her ability to handle the basic tasks of daily living, even control over bodily functions. He also knew how little hope there was for treatment.

It was heartbreaking. Hood confided in his old college friend, Eric Adelberger, on a camping trip to the Arctic National Wildlife Refuge in the early 2000s. For the first time in many years, Hood wasn't bending his friend's ear with monomaniacal intensity for systems biology. He started talking to a friend as a friend—the way he had as a young man.

"Lee knew it was Alzheimer's," Adelberger said. "We had both felt that if we ourselves were to get Alzheimer's, we didn't want to go on living as if we were vegetables. We talked a lot. It was an open conversation, like the old days."

Losing Valerie bit by bit was devastating. They'd met in high school. They both came from small-town Montana. They had shared values on the big things in a marriage—how to raise kids to be independent and to appreciate education. They loved the outdoors, traveling, and books. Valerie was vigorous and fit like her husband, sharing his zest for adventure. In their midfifties, they climbed Washington State's Mount Rainier, going all the way to the summit at 14,409 feet.

In Seattle, Valerie threw her energy into science education work, overseeing a small staff at the Institute for Systems Biology that revamped K–12 science teaching strategies around the state. She mingled with the business community and helped Hood raise money for science education reform.

Most importantly, she was his confidant and emotional rock. She provided perspective. At times, she kept his ego in check. "She challenged me on almost anything she felt I was wrong about, or arrogant about," Hood said. "She had no hesitation about doing that—sticking a pin in my puffed-up balloon. She was not a subservient wife."

When Hood needed help navigating tricky interpersonal conflicts at Caltech and the University of Washington, Valerie was there for him. Sometimes she sat in on important meetings to help her intellectually brilliant, but emotionally dim, husband better read people.

Now, at dinner with friends, she couldn't say much or follow the conversation. She became stressed in noisy situations. She struggled in unfamiliar places, unsure how to navigate the stairs when visiting her grandchildren in Alaska.

In 2005, Valerie's decline was noticeable but not debilitating. By 2010, it was unmistakable. Hood needed to become something he'd never been in his life: a caregiver for someone else. He didn't know what to do.

After Valerie was found wandering around dodgy parts of their Seattle neighborhood, Hood hired a part-time caregiver. But that was part-time. Hood needed to help his wife brush her teeth and take showers. Once, in the airport together, she lost control of her bowels. Hood shuddered at the memory. "I can't deal with that sort of thing," he said. For a man who fiercely valued independence, it was a hard concession to make. "He is so independent and so capable, he handles so many things himself," his daughter, Marqui, said. "But [he] does realize it takes a family." With much urging from friends and family, Hood hired full-time help.

More than anything else, Hood missed his wife's companion-ship. She was the only person he could truly open up to about the hard moments in life, such as when Caltech's biologists revolted, or when Bill Gates brushed off his systems biology dream. Hood was a resilient personality who could always pick himself up and charge ahead toward the next big goal. It was Logan who helped give him that strength. It shook him to the core when she wasn't there. He reflected:

> I'm psychologically pretty impervious to stress, in
> that I don't let it get me down. I'll tell you, the thing
> I feel most is an overwhelming sadness. I'm really

lonesome. Because I spent a lot of time with Valerie.
Those are the things I feel the most. ...

There was a lot of stress in leaving Caltech. She
knew all the details about that. There has been
stress at various times of my life, and Valerie was
always the one I could sit down and talk to. She
didn't always have answers, but at least she was
sympathetic and responsive. She's a very good
listener. She cared, she was always sympathetic,
and always an active problem solver. ...

For a variety of reasons, the last few years at ISB have
been really stressful. It's been very, very hard not to
have Valerie to talk to about all these kinds of things.
It lets you decompress. Talking verbally is a good way
to problem solve for me.

It also was hard, as a biologist, to be unable to help his wife.
Scientifically, so many questions remained unanswered at the time
of Valerie's diagnosis that it was hard even to know the most promising angle of attack. Hood noticed that some Alzheimer's patients
appeared to have even more perturbed molecular networks in the
liver than they do in the brain. It was a curious finding, which
researchers don't fully understand. "It's a really hard problem,"
Hood said. "What the networks are, how they get perturbed, which
ones initially get perturbed, and how you think about ways to avoid
that." Some scientists thought the disease might, at an early stage,
be more treatable with physical exercise and cognitive exercise.
Hood was interested in prevention. In the early 2010s, he sought to
find a group of signature proteins floating in the blood that might
be telltale early warning signs of Alzheimer's disease. But the work

never bore much fruit. Alzheimer's never became a primary focus of Hood's research.

Few at the Institute for Systems Biology knew the pressure Hood was under at home. "At some point he circled the wagons, and inside the circle was Lee and Valerie," Adelberger said. "He didn't open up to anyone outside." Instead, Hood turned to books for advice and solace. One was *Still Alice*, a novel by neuroscientist Lisa Genova, later made into a movie.

Hood said he didn't want to make people feel uncomfortable. "I have really good friends here, but I don't think they want to hear about my problems," Hood said. "I think they enjoy me because I'm interesting and I can talk about science. People don't want to talk about things like Alzheimer's. It hits too close to us."

Hood did see friends. He spent time at his retreat in the San Juan Islands with Tom Cable, a retired Seattle venture capitalist. Hood didn't have a TV in his own vacation home. When he wanted to watch a football game, he'd go over to his friend's house. The two men would unwind on weekends in front of the TV, cheering for the University of Washington Huskies and the Seattle Seahawks.

Hood made a new level of commitment to family. His travel schedule had been the biggest source of tension in his fifty years of marriage to Valerie. "She felt she was being cheated, and she felt the kids were being cheated," Hood said. "In retrospect, I can't argue with that at all. If I had it to do over again in life, I would have traveled a lot less." He vowed to visit his granddaughters every month. That meant he would fly from Seattle to Juneau, Alaska, to see Eran and his daughters one month, and the following month head in the other direction, to Los Angeles, to visit Marqui and her girls. Hood was proud of his children. Eran was an avid outdoorsman and a hydrologist who studied glaciers in Alaska. Marqui became an anti-discrimination lawyer. Hood also thoroughly enjoyed the time with his grandchildren. He directed his assistant in the fall of 2012 to book the dates on his calendar two years in advance, to make sure he scheduled work trips around these visits.

As Valerie slipped away, Hood clung to his work. He didn't realize that the Institute for Systems Biology was in danger of slipping through his fingers as well.

First, Ruedi Aebersold resigned. The cofounder and proteomics whiz left in 2004 to run a government lab in his native Switzerland. Officially, Aebersold and his wife said they wanted to be close to family. He had a generous offer. The Institute for Systems Biology had made it through the treacherous early years and wouldn't collapse if a cofounder left. Aebersold further claimed that if he and his wife were going to retire in their home country, then they needed to move before they got too old to qualify for Swiss retirement benefits. "There was nothing that drove me away," Aebersold insisted.

Unofficially, he had other reasons to leave. Aebersold saw an uncertain financial future for the institute, given its lack of an endowment. The Swiss government lab wouldn't continually be scrambling for grant funding.

Like every other accomplished scientist, Aebersold also had an ego. He'd taken a big career risk to cofound the institute. Although he made significant contributions, he was forever destined to operate in Hood's shadow. "When I tell people I worked at the Institute for Systems Biology, people say 'Oh, you worked at Lee Hood's institute,'" Aebersold said.

Louis Coffman was among many who urged Hood to take on the role of elder statesman. "I told him, 'Lee, the sun has shone brightly on you for good reason, for good cause. It will set. Let it shine on someone else, like Alan [Aderem] or Ruedi [Aebersold]. Somebody,'" Coffman said. Hood couldn't run the institute forever, and Aebersold was fifteen years younger. But as a reserved, introverted personality, Aebersold said he had no desire to succeed Hood and take on all the networking, fund-raising, public speaking, and media outreach that take time away from science. Hood wasn't interested in taking a back seat, either.

Losing a respected scientist and proven grant winner was a blow. Hood only briefly tried to persuade Aebersold to stay. He never brooded over losses. "Lee is a very pragmatic person," Aebersold said. Within minutes of hearing the bad news, Hood switched into problem-solving mode. "He said, 'How can we make this a win-win?' That's what we tried to do," Aebersold said. He was offered an honorific faculty title, and he remained cordial when informally discussing science with Hood and former colleagues at the institute. Hood was pleased. Aderem was floored by Hood's nonchalant reaction to the resignation of their cofounder.

"This is an opportunity," Hood said.

"How the fuck is this an opportunity?" Aderem shot back.

Aebersold's departure put more pressure on the institute's junior scientists. They were not as well positioned to win competitive federal grants, the institute's lifeblood. John Aitchison, one of the institute's first scientists, remembers whipping together an application for a systems biology center in two weeks. Normally, that process would take six months. "We were writing grant after grant," Aitchison said. "We were getting burned out."

Donors still weren't coming through with an endowment. The science was part of it. Systems biology was an abstract concept. You couldn't put it on a bumper sticker, like "A Cure for Cancer." It didn't pull on anyone's heartstrings. Even the most intellectually curious observers found it esoteric. "There were examples of [our] systems work in microbes or yeast," Aebersold said. Hearing that, "a person like Bill Gates or someone else with a lot of money will start to go to sleep. Why should they put $1 million into figuring out how yeast processes glucose? It's scientifically fantastically interesting, but is it something that's going to make a big impact on someone's life?" Donors couldn't see it.

The funding situation, along with Hood's age of sixty-six, stirred speculation about the future of the institute. Succession planning had been a casual topic of discussion almost from the institute's beginning in 2000. Naturally, by Aebersold's departure in 2004, the conversations had intensified. Some on the board argued that Hood was one-of-a-kind and it would be impossible to replace him. When Hood goes, the institute should shut down, they said. Others wanted the institute to endure. That meant grooming a successor, or multiple people who could shoulder various aspects of Hood's role.

Getting the timing right, as board member Perlmutter put it, was a "high-wire act." Organizations don't want to wait so long that they're caught flat-footed when the charismatic founder dies—as famously happened at Walt Disney Co. Organizations don't want to reach for the hook too early either. Merck forced out its legendary CEO, Roy Vagelos, in 1994 when he reached the mandatory retirement age of sixty-five. The company then struggled, while Vagelos remained productive for two more decades, making important contributions at a pair of successful biotech companies, Theravance and Regeneron Pharmaceuticals. But let the leader linger too long, and organizations tend to lose younger, ambitious talents who don't want to wait around for their turn.

Like Alan Aderem, the institute's remaining scientific cofounder.

"You can't sit here and be the crown prince forever," Perlmutter once told Aderem. The board, determined not to have another high-profile loss, urged Hood to create a succession path for Aderem. Aderem was given a seat on the board and elevated to the No. 2 position. The plan was clear. Aderem would continue running his own immunology lab, but he also would gain boardroom experience and handle internal management—which Hood was happy to unload. Hood would still lead the institute, craft the vision, and handle fund-raising and external partnership talks. Essentially,

Hood would be CEO. Aderem would be the chief operating officer and heir apparent.

Aderem waited. And waited. When would Hood step aside? When I first asked him about retirement in 2002, when he was sixty-four, Hood said he planned to run the institute four to six more years. It was a reasonable thing to say to a local newspaper reporter. But by age seventy, Hood had no intention of riding off quietly into the sunset. He felt great physically. He wanted to keep going. He worked hard to dispel any notions people might have that his best days were behind him. "I'm doing the most ambitious things, by far, that I've ever done in my career," Hood told me days before his seventieth birthday. "Right now." His whole identity was wrapped up in his work, thinking about the future, being on the leading edge. The institute was an expression of everything he held dear. It was an expression of his values in science, education, business, and the transfer of knowledge to society. He had fought his entire career for the power to operate on his own terms. Finally, he had it. He couldn't let go.

The man's schedule remained relentless.

For his seventieth birthday celebration, at the swanky W Hotel downtown, more than two hundred people attended—including Irv Weissman of Stanford, George Church of Harvard, and Lee Silver of Princeton. Hood's staff showed a series of tributes they'd videotaped. Most of them focused on Hood's stamina. Software mogul Bill Gates, in his video tribute, said:

> Congratulations on your seventieth birthday. I
> can hardly believe you're that old. Your energy is
> amazing. I know in the next decade you're going to
> achieve even more fantastic things. I feel sure some
> of your breakthroughs will make a difference to
> people all over the world, and I'll do my best to make

sure they get deployed. Keep up the good work. ...
I'm a big believer in the great work Lee is driving at
the Institute for Systems Biology.

Hood worked an average of eighty-four hours a week—twelve
hours a day, seven days a week—all the way through 2012, the year
he turned seventy-four. He took home about $382,000 in salary that
year, about $100,000 more than the next highest-paid officer. He
was on the road almost every week, routinely logging more than fifty
trips a year. "He's absolutely unstoppable," Aderem said. "The guy's
a maniac."

"I remember saying, 'Lee, people come here to the institute to
rub shoulders with you,'" Coffman said. "I remember once he was
going off to a trip to Galveston [Texas]. I remember saying, 'What on
earth could possibly be going on in Galveston that's more important
than you being here?'... He was just going to give talks. He loved
being in the spotlight."

With Hood constantly on the road, the all-for-one, one-for-all
start-up culture faded. The institute developed a revolving door,
especially on the administrative side. Directors and vice presidents
of fund-raising, operations, finance, and business development all
came and went. Many arrived with sterling credentials, eager to
work with a legend. They quickly found they couldn't work with
him.

For example, when Larry Herron, then vice president of fund-
raising, sensed a donor was about to bite, he would lay out a follow-
up plan designed to reel in the check within six weeks. He knew how
much hand-holding philanthropists required. Hood usually thought
that was unnecessary. He didn't want to make multiple follow-up
phone calls, especially for a small amount of money.

Herron also wanted Hood to create a long-range strategic plan.
Many prospective donors liked Hood's message, Herron said, but

they were uncertain about the longevity of the institute, without a succession plan or even a well-written ten-year to twenty-year organizational plan. Hood felt a written plan would create constraints, making it harder to pivot to the next hot thing, whatever it might be. Nimbleness served Hood well scientifically. But potential donors interpreted it as being wishy-washy, lacking in commitment to the causes they held dear. "I remember telling him, these people are in demand all the time because they're rich," Herron said. "I'd say, 'You're one of the more intriguing people they'll meet, but you're hardly the only one. Why should they invest in a smart guy in Seattle who's running an institute and can't tell them what he's going to be doing in five years?' We had those conversations many times."

Despite Hood's general kindness, generosity, and loyalty to his employees, his tone deafness for people's feelings was often the final straw. Herron recalled once being savagely criticized in a consultant's report. The report said that fund-raising at the institute was a disaster and Herron should be fired. Hood, apparently without reading the report, distributed it among Herron's administrative peers for review. Alarmed, Herron called the boss while he was traveling in China. Herron was proud of some fund-raising wins that occurred on his watch, particularly a $6 million donation from Bill Bowes, the California venture capitalist. He wanted to know if Hood agreed with the consultant's damning summary. No, Hood didn't. The boss told Herron he was doing well and should stay. Still, embarrassed by the episode and frustrated by internal conflicts, Herron quit a few months later.

The institute continued to lose money—sometimes millions of dollars a year—despite respectable grant support. With no big-name donor, Hood continued trying to forge "strategic partnerships" with foreign governments. His pitch went something like this: give the Institute for Systems Biology a lot of money and we'll help create a biotech industry in your country—one that puts your country on the leading edge. Painfully, Hood had spent a lot of time and money in the early years of the institute chasing deals with various governments. Nothing had materialized. Finally, after seven years of talks, the government of Luxembourg bit.

It was a godsend. Then it almost brought down his entire institute.

Luxembourg, a tiny European nation few Americans could find on a map, was an odd place for Hood to emerge victorious. It was one of the few countries on earth he'd never visited. He didn't know any scientists there. Luxembourg is a landlocked country smaller than the state of Rhode Island, bordered by Belgium, France, and Germany. The head of state is a grand duke. It's the wealthiest country in the world based on per capita gross domestic product, as tallied by the International Monetary Fund, but the population is less than that of Seattle. The economy was dominated in the first half of the twentieth century by steel production, before Luxembourg diversified into financial services. By the time Hood was selling his concept of strategic partnerships and economic development, Luxembourg was looking to diversify again. An acquaintance from the consulting firm PricewaterhouseCoopers helped broker a meeting. Money was no object. "You can ask for whatever you like," the consultant said.

Hood asked for $100 million over five years. It was time to be bold.

The University of Luxembourg was then only four years old. In biology, "they had no idea what they were doing," Hood said. Advisers to Hood were skeptical, given the string of failed partnerships elsewhere. "If you succeed in this, I'll believe in the tooth fairy," one institute board member said.

Unlike other places where decisions can die in committee, Hood said, he needed to persuade only one man—Jeannot Krecké, the country's economic minister. Partly on the advice of PwC, Luxembourg found a couple of other top scientists, Lee Hartwell at the Seattle-based Fred Hutchinson Cancer Research Center and Jeff Trent at Arizona's Translational Genomics Research Institute—to join the proposed collaboration.

There was no formal scientific review. The idea was for Hood and the Institute for Systems Biology to help establish a bioscience research center in Luxembourg that could spin off companies. Personalized medicine was the big idea.

Luxembourg was particularly intrigued by one development from the Institute for Systems Biology. It was a test that promised to detect lung cancer at its earliest and most treatable stage. The test needed only a small sample of blood to hunt for traces of signature proteins thrown off by tumor cells in the bloodstream. An initial attempt at commercialization in the United States had failed; the measuring devices weren't sensitive enough. By 2008, the underlying technology had improved. Hood was eager to try again, and Luxembourg was ready to get behind it.

Luxembourg would invest in a company, Integrated Diagnostics, to commercialize the test. Hood rhapsodized that this application of personalized medicine would be the crowning achievement of his career. The data supporting the venture were thin, but it hardly mattered. Luxembourg had faith in Hood and his Midas touch. "Their whole motivation was about 90 percent economics and 10 percent science," Hood said.

Talks moved fast. Hood negotiated without keeping his administrative staff in the loop, and the deal was basically complete by the time Hood presented the contracts internally. Confusion swirled. Hood didn't see why this might bother his team. Any details that needed working out seemed minor.

In June 2008, Hood made the dream announcement. Luxembourg pledged $100 million over five years to the Institute for Systems Biology. Maybe Hood wouldn't raise a philanthropic endowment, but this was the next best thing: a big pot of money with no strings attached. Within months, the financial crisis whipsawed the global economy, but Hood was riding high. When the Luxembourg deal was struck, the institute had fourteen faculty members, 230 employees, and an annual budget of $35 million. The budget leaped ahead to $55 million in 2009, enabling Hood to hire dozens more employees and begin outfitting a building twice the size, with room for 330 employees. People at the institute scrambled to figure out what to do with the loot. "It was like a moderately sized python engorging an extraordinarily large pig," Herron said.

Almost immediately, infighting over the money turned nasty.

David Galas was put in charge of managing the Luxembourg collaboration. Hood had known Galas since the early days of the Human Genome Project. Back then, Hood was pushing the project, the Department of Energy was in the driver's seat, and Galas was a well-placed official in the department. Galas went on to become the CEO of Darwin Molecular, the start-up cofounded by Hood that did well for investors Bill Gates and Paul Allen. When the company was sold, Hood hired Galas as a faculty member at the Institute for Systems Biology. The two had a close relationship—a friendship—for more than twenty years.

Luxembourg officials, ever mindful of economic development, wanted not just to support basic research at the Institute for Systems Biology—they wanted an ownership stake in some of the business applications that grew out of the research. That's where things got murky for Hood and Galas. Both would serve as cofounders and officers of the for-profit spin-off company, Integrated Diagnostics. Simultaneously, both had responsibilities to the institute, which was a nonprofit. Both the nonprofit and the for-profit depended on Luxembourg cash. Hood and Galas were the ones in contact with the Luxembourg benefactors. People within the institute wondered: How could they strike a fair deal for the company *and* the institute when they were negotiating with themselves?

Hood shrugged off the objections, as he did many times throughout his career, saying the conflict was easily manageable, and he meant to do well for both sides. But a promise wasn't enough. Even if the conflict of interest were effectively managed, there were legal governance controls and accounting issues that couldn't be taken lightly. The nonprofit institute had loaned the for-profit company $288,000 during 2008, according to federal tax records. The company had physically set up shop within the nonprofit institute. It wasn't entirely clear to many inside the institute how the money, say, for office rent, was being controlled, accounted for, and paid back to the institute.

Handled with great care and attention to detail, the situation could be managed. Handled cavalierly, it could potentially jeopardize the institute's tax-exempt nonprofit status. That would be the death knell for an organization that lives on federal research grants.

The blurring of lines between nonprofit and for-profit, the confusion over who knew what, and who had the power to do what, raised concerns at the board level. Three respected trustees on the institute's board—Steve Clifford, Jon Runstad, and Hank Riggs—resigned in rapid succession. Around this time, Perlmutter scaled back his involvement after becoming chairman of the board of his alma mater, Reed College. Perlmutter said he believed the conflicts of interest were manageable, if treated with care.

But Hood wasn't careful. He had always seen big pots of money as "fungible" for any purpose he saw fit. He had always run into conflict with careful financial managers, had always made them nervous about audits. Hood could never bring himself to think in "columns and rows" on spreadsheets, as Perlmutter put it. Hood wasn't paying attention to how Galas managed the money, either, or the simmering internal conflicts over the money.

"He's a true visionary. He sees the future very clearly. Sometimes he's not so good in the present," Perlmutter said.

The board resignations were kept quiet; few on staff knew. What couldn't be hidden was that a culture of team spirit had been eroded by fear, suspicion, and the emergence of rival factions. Aderem clashed bitterly with Galas over the money. Aderem, who had raised $97 million through federal research grants during his decade at the institute, was so nervous about how the Luxembourg money was being handled that he vowed not to take any of it (although his lab moved into the nice new building along with everyone else).

Galas worked closely on the partnership along with his wife, Diane Isonaka, and another institute scientist, Joseph Nadeau. There were no peer reviews and no clear system for prioritizing which projects got funded and which didn't. Some good research did receive funding, including important work by Hood's team to sequence the genomes of a family with a rare disease called Miller syndrome. But many scientists in-house were concerned at what they saw as an overly subjective funding process. The thing looked like a private slush fund, at least to those in the Aderem camp.

Months went by before Hood began to sense that something was amiss. A longtime scientific ally, Eric Davidson at Caltech, thought Hood suffered from a certain gullibility and naïveté about human nature. Hood had a tendency to remain loyal, he said, to people who ultimately stabbed him in the back. Each time, he didn't see it coming. "Some of this is from the cultural and social history of the Midwest," Davidson said, referring loosely to Hood's small-town upbringing in the northern plains of Montana. "Valerie is the same way. Out of that grows a lot of do-gooder attitudes. If you want to take a broader point of view, how often do people who are trying to do good end up being betrayed by the people they are trying to do good for? It happens all the time. He has that streak. It's caused him untold troubles."

At the suggestion of his finance chief, Jim Ladd, Hood asked for an independent review by an experienced organizational psychologist.

The review went on for more than a year. The psychologist found an institute split into factions, built-up animosity, impatience, and thwarted ambitions of those who saw themselves as the rightful successors to Hood. The leader had serious blind spots about what was happening around him.

With the help of the organizational psychologist, Hood began to look in the mirror. He surprised many close colleagues by showing some humility. He took the unusual step of accepting responsibility for his mistakes, acknowledging that his actions, or lack of actions, helped create a combustible situation. He admitted he should have been "more cynical," seen the personnel problems coming and nipped them in the bud. "That was a major failing on my part. No question," Hood said. But it was too late.

Alan Aderem had had enough. Not only was he disturbed by the Luxembourg partnership, but he could clearly see that Hood— even at age seventy-two—wasn't going anywhere. In 2011, Aderem accepted an offer to become the president of the nonprofit across the street: Seattle Biomedical Research Institute, later renamed the Center for Infectious Disease Research.

Aderem's exit was a body blow. Aderem took forty people with him, along with a $7 million grant from the Bill & Melinda Gates Foundation. Aderem said he was still a friend and admirer of Hood. But now he could advance his own ideas on how to apply systems biology. Aderem, a native of South Africa, wanted to use the computational tools of systems biology for global health purposes, like vaccines for the developing world. Hood was wedded to his broader concept of personalized medicine.

"Alan has wanted to assume a leadership role for some time," Hood said, "and I'm not ready to step down."

But Aderem and his team's departure was only the beginning.

Within a year, half of Hood's administrative staff was in open revolt, threatening to resign unless he fired Galas, Isonaka, and

Nadeau over their handling of the Luxembourg money and some angry outbursts over petty issues, like the paint scheme at the new office. A board member urged an internal investigation, which revealed a series of e-mails that Hood interpreted as insubordinate. Hood dreaded firing anyone—especially friends whose wedding he had attended—but he couldn't ignore the open revolt. The situation was so heated that lawyers stood at Hood's side as witnesses during the trio of terminations, as personal belongings were picked up and hauled out. The institute was thunderstruck by the chain of events.

The Luxembourg collaboration wasn't renewed. The partnership's scientific director, Rudi Balling, said Luxembourg officials had gotten what they wanted in the five-year term. With the Institute for Systems Biology's help, the country had set up its own systems biology center with thirteen researchers and 180 employees. Nearly two hundred published papers described the scientific work in Luxembourg. The new center no longer needed mentoring; it could stand on its own. Despite the tensions, the partnership had been successful.

It wasn't until this point, in the fall of 2012, that Hood voluntarily brought up succession planning. He was seventy-four.

In seeking to replace high-caliber scientists like Aebersold and Aderem, however, Hood was in a tough spot. By this time, Hood's reputation as a self-absorbed demigod tended to repel the best scientists. Weissman, the Stanford biologist and Hood's lifelong friend, was among those who said Hood began to attract second-rate people. Working with Hood meant sublimating one's own ego, taking a back seat. Many top people weren't interested. Hood could recruit scientists with certain kinds of skills, but no one with the whole package. The institute was struggling not only with the lack of a $100 million endowment, but with an annual deficit of about $5 million—also not attractive. During a period of financial strain, and some uncertainty about the institute's long-term prospects, Hood had merger talks with a couple universities. He wanted to see if they would essentially take over the institute, financially support it, and keep it in Seattle. That never got beyond the talking stage. Getting a world-class successor at the freestanding institute, he said, hinged on his ability to make the job attractive. That was going to be difficult.

Hood, of course, was as excited as ever about his vision for personalized medicine. That always got him going in the morning. Systems biology laid the building blocks for "predictive, preventive, personalized" medicine. By studying whole networks of genes and cells, physicians could see when someone was at risk of disease and starting to fall out of molecular equilibrium. They could act much earlier, before symptoms became too severe. Treatments could be tailored for the precise molecular defect in the individual. Hood later added a fourth p, for "participatory"—"P4" medicine, for short. Public participation was essential. Large numbers of patients would need to volunteer to contribute blood, saliva, or tissue samples so that scientists could map out human biological networks and then compare them from one person to the next.

Hood delivered his gospel-like message via presentations that contained little hard data. Many biology peers dismissed it as salesmanship and sloganeering. Hood struck a few modest partnerships, including one with Ohio State University. He didn't disrupt the American health-care system. Mark Davis, the Stanford biologist and former student whom Hood tried to chase down in the T cell receptor race, in early 2013 said:

> The P4 thing, it's such bullshit. It's not helpful. It pains
> me to say that, because I admired his intellect and
> worked with him in a great period of his professional
> life when he made his reputation. But in recent times,
> I think he maybe got too impressed with himself or
> found he could get by being this sort of vague visionary.

Most serious scientists don't pay much attention
to him anymore. He doesn't say anything we don't
already know.

Health economists, health insurers, and policy makers also were unimpressed, for different reasons. Personalized medicine promised to be expensive, just as policy makers were seeking to corral health-care costs. Hospitals were usually unwilling to share samples with scientists, partly out of concerns about legal liability and patient privacy. They were in the habit of tossing out samples in the medical waste. It was hard to imagine they'd suddenly start encouraging people to participate based on some futuristic notion of the greater good.

Hood, ever the optimist, switched gears.

It was the summer of 2013, when the lucrative Luxembourg contract was due to expire. President Barack Obama had recently presented him the National Medal of Science, with his family there to share in the moment. The institute's culture had stabilized. The time was right to reimagine the future of the Institute for Systems Biology.

He would embark on a breathtakingly ambitious wellness study.

The idea was vintage Hood: inspiring in scope, far beyond the capabilities of existing technologies, and way too expensive for any one person or agency to bankroll by itself. The idea was to enroll one hundred thousand people, sequence their genomes and follow them for twenty years. The ultimate price tag of such a project, to the extent it could be estimated at all, would surely run into the tens of billions of dollars. As usual, he started with an intoxicating vision. He could fill in the details on the fly.

He dubbed it "the Framingham study of wellness." The project, drawing inspiration from the monumental six-decade heart study of health-care workers in a small Massachusetts town, had an even more audacious aim. The hundred thousand people would be asked to give blood, saliva, and stool samples every three months to track a litany of other molecular markers that fluctuate over time and might signal the onset of disease. Patients would be informed of any warning signs so they could change their behavior and see what difference it made.

Yet again, this was to be the big achievement of Hood's life, his true legacy. The first year alone, with just 108 healthy volunteers, was quite doable. It was estimated to cost $2.5 million. Hood had no firm commitments to finance the rest of the study. Yet.

Hood hadn't lost his entrepreneurial streak. While raising money for the rigorous scientific wellness study, Hood simultaneously worked to start a for-profit company that would do a cheaper, more modest form of wellness monitoring for consumers. That sort of thing was beyond the pale for many peers in biology, who consider wellness to be in the realm of the hucksterism seen on late-night TV infomercials. Hood risked a late-career embarrassment akin to Linus Pauling's obsession with vitamin C. Hood knew it. "Wellness is kind of a schlocky, semicommercial, questionable-reputation kind of thing, and it's been around for some time," he said. "We plan to insert quantitative scientific analysis into it and provide it with real measurements."

Hood himself joined the 108 original volunteers. Within months, he declared it was both a scientific and personal revelation.

Through the early years of genomics, Hood never bothered to get his own genome sequenced. Scientists such as James Watson and Craig Venter did. By waiting for the technology to improve, Hood knew his genome sequence would be cheaper and more accurate. The information would be interesting personally. But how could it advance science? The way, he realized, was not to look at his genetic code in isolation, but to compare it with that of three generations of Hood family members. Because family members share so much

DNA, researchers are able to compare the genomes and focus on differences that count. Hood's lab had already successfully sequenced the genomes of a family of four with the rare Miller syndrome. Researchers were able to identify the genetic cause of the disease: exciting. Equally important, the work, published in 2010, was the first to demonstrate the usefulness of sequencing a whole family. This was a stirring finding. It showed how by gathering interesting data, you could turn it into practical medical knowledge.

In August 2013, Hood gathered about thirty of his family members at the Montana ranch near the Bitterroot River. All gave blood samples that could be sequenced in a large study. By looking comparatively at families with similar genomes, Hood was hoping to narrow the search for medically interesting variations, including some that get passed down through inheritance. Complete Genomics, a sequencing company in the San Francisco Bay Area on whose board Hood sat, agreed to sequence the genomes for $1,500 apiece. To protect family members' privacy, Hood kept the only master list of names identifying the samples. They sit on a piece of paper in a secure location, to avoid any possibility of computer hacking, Hood says. "I'd like to write a series of papers on this as a classic study of a family," he said.

So Hood had his genome. He added the battery of the tests from the wellness study. Hood found he had deficiencies in vitamins C, B1, and B12. These were at least partly caused by genetic mutations that made it more difficult for him to get the recommended daily allowance. He was able to correct the deficiencies by taking high-dose supplements. He learned a bit about the genetic basis for his risk of cardiovascular disease (his great-grandfather, grandfather, and father all died of heart attacks before the age of seventy). Based on the flood of wellness data he gathered over time, Hood made some significant lifestyle changes. He continued to do his set of more than one hundred push-ups every morning, along with sit-ups and stretching exercises. Now he increased his exercise routine from twenty minutes to about seventy-five minutes a day. That included time on weight machines and treadmills. He greatly reduced carbohydrates from his diet, and he all but quit eating dessert when going out for dinner. He started sleeping seven hours a night, instead of his usual five to six hours.

The results, Hood said in the fall of 2014, were spectacular. On the verge of turning seventy-six, he said he felt the best he had in years. As if he'd found his own fountain of youth. He shed thirteen pounds, getting down to 180, and wanted to drop about five more to get back to his college football–playing weight.

Hood was so fired up about the potential for this sort of quantitative, data-driven approach to wellness that he entertained notions of making a pitch to Congress to fund a ten-year wellness project that would dwarf the Human Genome Project. "The preliminary analyses are just revolutionary," Hood said. "I've gone around the world and country. Everywhere I've gone there is enormous enthusiasm for this. I'm thinking very seriously of taking this project to Congress and making it a second genome project. I could go to Congress and say, 'Look, I can promise you more with this project than I ever could with the genome project.'" Such grand notions ignored the partisan gridlock in Congress. Not to mention the long-term trend toward cutting biomedical research budgets rather than increasing them.

Hood also butted up against wide and deep skepticism among biologists, physicians, statisticians, and regulators. They widely believed that much of the wellness data being gathered was trivial or meaningless. They wanted to see clinical trial results.

The established players had reason to be skeptical. Sure, an individual could have beneficial experiences based on wellness data, but they could get that benefit from any number of changes: eating more fruits and vegetables, or sleeping eight hours a night. Without rigorous controls in a study, scientists couldn't say for sure what effect any of those changes might be having. In testing new drugs, scientists have learned that blinded clinical trials can wipe away most of the bias from physicians and patients in a study. Nobody knows who is getting the new health intervention being tested, whether it's a drug or something else. Patients are randomly assigned to a new drug or a placebo to further reduce the effect of other factors and help researchers zero in specifically on what the new intervention is doing. Conclusions are reached based only on the results of many, many people who got either a new drug or a placebo. That's the gold standard of medical evidence that physicians use to make decisions.

Doctors had spent years moving away from anecdotes and gut feel and toward this more statistically rigorous form of evidence-based medicine. The problem with those randomized clinical trials, for Hood, was that they dealt in medians and averages. They couldn't say precisely how a given treatment would affect an individual. Some people might get tremendous benefit from a new treatment, while others wouldn't be helped at all. And yet the focus would be on the median or the average. Evidence-based medicine was certainly better than gut feel and hunches, Hood agreed. But still in many respects, it amounted to generalized recommendations and prescriptions that weren't well tailored to individuals.

To Hood, that wasn't good enough. Every individual patient was genetically unique. If a study with three thousand people was N of 3,000, this would be N of 1. Traditionalists scoffed. Medicine had a long history of getting carried away with excitement over anecdotal successes that couldn't be repeated. But Hood argued, and some scientists agreed with him, that the new data-heavy form of biology made it possible for individual patients to serve as their own study control, to filter out biases. For example, the physician would take a wide range of measurements at the beginning of a study, perturb the patient's organ system with a new drug or supplement, and see how it changed the body's molecular profile—before symptoms emerged. The new biology produced huge amounts of data, making it possible to measure changes at the molecular level and to connect those changes with observable, and clinically meaningful results, like a reduction in disease symptoms.

Hood chided skeptics in the medical profession, the pharmaceutical industry, and the Food and Drug Administration for being slow to embrace the new tools of science and the emerging forms of data analysis. "They've lived the old-fashioned way," Hood said. "They need to start thinking in new ways. N-of-1 experiments are going to become important. This is enormously controversial. But I guarantee you, I've been right in the past, and I'm going to be right on this one." Still, the big federal grants didn't materialize.

The for-profit wellness monitoring company found a friendlier reception. What if you could package a complete genome along with other changing quantitative measurements of health—from blood, saliva, and stool—and throw in some wellness "coaches" to help consumers interpret the data and make appropriate lifestyle changes? So the US government didn't want another top-down, Human Genome Project–type revolution. Maybe this could be a bottom-up, consumer-driven revolution.

Maveron, the venture capital firm cofounded by Starbucks mogul Howard Schultz, had shown special aptitude for investing in start-ups that tapped into consumer attitudes and desires. They were interested. Other venture capitalists were, too. By July 2015, the company, called Arivale, raised $36 million to bring "scientific wellness" to the masses. For $2,000 apiece, early adopters could get the same program that Hood said had changed his state of wellness. At that price, the company lost money on each test. Within a few months, Arivale raised the price to $4,000 a person. Consumers who wanted the test would have to pay out of pocket—insurers wouldn't cover it.

Hood wasn't the only major scientist gravitating toward wellness in the 2012–13 period. He couldn't help but wince when he saw other high-profile biologists beat him to the punch with big ideas. Google offered former Genentech CEO Arthur Levinson, a distinguished scientist, a blank check to pursue life-extension research at a company called Calico. Craig Venter raised $70 million in March 2014 for a start-up called Human Longevity—double what Hood's company raised.

Retirement? Hood continued to keep up a schedule that would wear out men thirty years younger. By February 2013, Hood was serving as a board member for thirteen for-profit and nonprofit organizations. He simultaneously sat on scientific advisory boards for sixteen other companies and organizations. ("I learned almost as much from my interactions with companies as I did from interactions I had at scientific meetings," Hood later told a reporter at the *Scientist*.) At age seventy-five, his curriculum vitae—which included all of his peer-reviewed papers, awards, guest lectures, and company affiliations—stretched to an astonishing seventy-five pages. His contemporary Irv Weissman's was thirty-three pages at that time.

Privately, though, Hood started to feel the effects of aging. He talked with friends about what it's like to slow down, to feel more aches and pains. His posture stooped. His chin would often tilt down, giving him the appearance of a battering ram. A hand tremor, a minor nuisance for years, became more noticeable. Although he had a reputation for taking catnaps during seminars, Hood appeared to sleep longer and more deeply during science talks later in his career. One scientific collaborator said Hood once fell asleep in an important meeting with high-powered pharmaceutical executives.

Mortality began to loom.

There was his wife's slow fade. By early 2014, after many talks within the family, Hood finally agreed: it was time for Valerie to go to a home where she could get round-the-clock care. The family picked a comfortable assisted-living facility in Seattle's Madison Valley neighborhood, about a fifteen-minute drive from Hood's condo. "I have all sorts of feelings," Hood said shortly afterward. "One, I feel guilty that I'm not going to take care of her and provide a home. Two, we've been married fifty years. I haven't lived by myself for a long time. I'm going to spend a lot of time visiting her." To recognize his wife's long-standing commitment to science education, Hood made sure that his institute's Center for Inquiry Science was renamed the Logan Center in her honor. An annual community fund-raiser for the cause would be done in her name, not his.

Hood's father had died at sixty-nine of cardiovascular disease. He was deeply shaken when, one day in the fall of 2015, his long-time Caltech peer and friend, Eric Davidson, died suddenly at age seventy-eight.

For all that, Hood had some of his happiest moments in these years with his grandchildren. Although he didn't climb mountains anymore, he was in good enough shape to get down on the floor and play. Marqui, who had some difficult times with her dad as a teenager, became closer to him. She spoke with joy about how the world-class scientist in his seventies could show up at her house and forget about everything for a few hours to be a grandfather.

> When I was small, one of my strongest memories was playing tickle monster with him. He'd get on his hands and knees and chase us until we'd become hysterical. A lot of memories are so salient to me, because I have the experience of watching him do this with my kids [ages three and five] now.

> He has an element of being able to be in the moment and have immense enthusiasm for what's going on in that moment. ... It makes him a great grandfather.

The Arivale-inspired changes also got Hood feeling more optimistic about his longevity. Nearing his seventy-seventh birthday, Hood said he felt so good that he imagined staying active well into his nineties. A year earlier, he had pegged it at eighty. He looked to Francis Crick as a model of how a scientist could remain productive to the end. Crick—the codiscoverer of the DNA double helix and one of history's most influential biologists—spent his final years studying neuroscience at the Salk Institute in San Diego. There he remained a creative and influential thinker into his eighties. Hood had visited Crick in San Diego just a couple months before he died in 2004, and the two had a stimulating conversation about how systems biology might help crack some of the tough problems in neuroscience. "The only thing you could notice is that after a few hours, he got tired," Hood said.

Most longtime friends and colleagues couldn't imagine Hood ever riding into the sunset gracefully. "He'll die on some trip looking for accolades," said Eric Adelberger. "From my point of view, it would be sad. It would be nice if he could somehow become more open to more things. Appreciate his life. Be reflective."

After seeing many succession plans fizzle, Aderem said Hood would simply work on science and business until he physically couldn't. The Institute for Systems Biology didn't grow into a powerhouse with twenty-five faculty-led research groups, as Hood once envisioned. By early 2016, it had nine faculty-led research teams. Fund-raising was a constant battle in an era of demoralizing federal research spending cuts. The institute subleased some excess space at its South Lake Union headquarters to a branch of Northeastern University.

Then, Seattle real-estate mogul Dave Sabey, a director of the institute, had an idea. He was friendly with Rod Hochman, the CEO of Providence Health & Services in Seattle. Providence was also a health-care network, like Kaiser Permanente. It operated by collecting insurance premiums from members and providing care through thirty-four hospitals and six hundred physician clinics spread across five states: Alaska, California, Montana, Oregon, and Washington. The Providence network had eighty-two thousand employees—doctors, nurses, physician assistants. It was the third-largest US nonprofit health-care system.

Sabey knew that Providence had an interest in making a splash in the genomics and personalized medicine world. He brokered a meeting with Hood.

Providence was an unusual match for Hood at first glance. It was a fully integrated, nonprofit Catholic health-care system. It had a far-flung group of donors deeply committed to its mission of providing top-quality health care, including to indigent people and those in rural areas. The organization had no scientists, and few connections to Hood's realm.

What Providence did have was money. The Patient Protection and Affordable Care Act—commonly known as Obamacare—shouldered many expenses for the care of uninsured people. Before, Providence had paid the bills for uninsured health care out of charity care accounts. Now revenue surged 16 percent in a year, and it had $5.8 billion in cash reserves.

Flush with cash, Providence needed to figure out how to put its money to productive use. As Obamacare tightened the screws on reimbursement rates, hospital chains merged to achieve greater economies of scale. In this new world order, the hospital chains sometimes marketed themselves to the public and to donors based on factors other than their quality of care. Some pitched their access to cutting-edge technologies and research prowess.

In Seattle, the University of Washington Medical Center did everything from research in the petri dish all the way through patient care. Virginia Mason Medical Center, another major Seattle hospital, had a long affiliation with the Benaroya Research Institute.

There, across the street, scientists could do cutting-edge research and collaborate with physicians who treated patients. Here was a chance for Providence to brag about the same.

Providence could tap its powerhouse network of donors in five states and offer them something tantalizing and new: the possibility of helping to realize, through research, a grand vision for personalized medicine.

Eventually, for Hood, Providence was a match made in heaven. The financial strength of Providence was almost irresistible. It could provide the sort of financial stability he could never achieve on his own. "For fifteen years, I have struggled every single year to make payroll," Hood told me in early 2016. No longer would the institute live in fear that it would dissolve.

Equally important, Hood hit it off with Hochman, a physician by training, not a finance guy or lawyer like some hospital CEOs.

They struck a deal. Providence essentially acquired the Institute for Systems Biology. Financial terms weren't disclosed. The arrangement, technically an "affiliation agreement," allowed the institute to remain a freestanding 501(c)(3) nonprofit organization, with its own board, but with financial results consolidated under the umbrella of Providence. Hood would become a Providence employee, with the title of chief science officer for the network. The institute, as always, would have to continue to recruit strong scientists who could find grant support for their research. Providence would provide the cushion where grants fell short. It also would consider funding projects that were too speculative for peer reviewers at the NIH.

And Hood would have not one boss, but two.

Hood said he could live with that at this stage in his career. They assured him that he would retain his operating autonomy and that bureaucracy would be kept to a minimum. "They have their red tape. Every large organization does," Hood said, reflecting on the deal a few weeks after it was announced. "I think I've wired my reporting so I can deal with it." He spoke of spending more time on science and less time traveling the world in a fruitless search for partnerships.

Overnight, Providence delivered the financial ballast that made the institute more attractive to potential scientific recruits. Hood's first wish: double the institute faculty, which he had wanted to do for years but couldn't afford.

The Arivale program also got a boost. Providence agreed to study its own health-care workers to see what secrets the genome, blood, saliva, and stool held about improving their health.

And Hood turned again to Alzheimer's. He wanted to look at several hundred genetic variations, to determine what puts people at risk of developing Alzheimer's. Then he would track people over time, evaluate a variety of changes to their behavior and lifestyle, and see whether symptoms could be delayed. Such ideas had been sitting on the shelf at the institute, waiting for funding.

Aderem, the cofounder of the Institute for Systems Biology, marveled at how the man remained a fountain of ideas—some good, some not so much—into his late seventies. Succession planning was not among them.

With his intense gaze, Hood looked ahead. He brimmed with enthusiasm: exciting scientific ideas awaited. The Hood legacy would not be charted or planned. Hood would continue to follow his scientific instincts wherever they might lead. Younger scientists were welcome to join him on the journey. There just wouldn't be a carefully plotted passing of any torch.

"Lee's succession plan," Aderem said, "is death."

Epilogue

―――――

Legacy of a
Scientific Entrepreneur

Lee Hood grew up in an optimistic time and place in American history. Adults in his small-town Montana community instilled that most precious belief: that anything is possible for someone with talent and a work ethic. Hood accepted it, believing he was destined for greatness.

His parents expected as much. They passed on two core values, education and independence. Hood's father—brilliant but unable to display pride or affection—instilled in Hood a deeper driving force: a lifelong quest for respect and validation from others.

Teachers encouraged the boy, urging him to push himself beyond his remote western plains upbringing. The tiny community treated him like the chosen one, as valedictorian and quarterback of an undefeated high school football team. Through a fortuitous twist of fate, he met a caring teacher who steered him toward the great minds of Caltech. There, he drove himself ferociously to compete with students who were more gifted. Insecure about whether he belonged, he projected an outward confidence that served as a defensive shield. He used that shield as a battering ram throughout his career, whenever he encountered skeptics.

Hood channeled all of his ingenuity and drive, in graduate school at Caltech, into hard immunology problems. One tough biological puzzle led to another as he advanced. Unsatisfied with the standard lab tools and techniques of his day, he resolved to develop better ones. That prompted him to take a dual-path career track, in which he spent part of his time on science and the other part developing enabling technologies for science. It was unorthodox and controversial, but he persisted.

By his seventh decade, the man had made a singular mark.

Often, scientists develop a narrow specialty and reach their creative peak early in their professional careers. Instead, Hood drove himself with uncommon intensity to stay on the cutting edge. That constant reinvention, that striver's mind-set, enabled him to hold on to a leadership position in biology for more than forty years. He blazed trails in four major disciplines: immunology, genomics, systems biology, and computational biology. His name appeared on more than 750 published scientific papers. Luck, no question, played a part. Hood's career took off in the 1960s, a time of ample government funding for science. He was in his prime when, in the early 1980s, the world woke up to the possibilities for biotechnology. Hood's wide range of accomplishments put him in an elite category. He is one of just twenty-two people who have been elected by peers into all three US national academies: the National Academy of Sciences, the National Academy of Engineering, and the Institute of Medicine.

What did Hood accomplish? Assessments gravitate toward extremes.

To his admirers, Hood is one of the greatest visionaries for biology in history. They always use that word: *visionary*. But Hood's career wasn't all about grand notions of the future. It was firmly rooted from the start in meticulous, painstaking research in immunology. Specifically, Hood led teams that made a series of discoveries into how the adaptive immune system uses genetic rearrangements to defend the body from an amazingly diverse and relentless parade of invading pathogens. The research, spread over two decades, forced scientists to rethink basic assumptions about evolution and genetics. That hard-won knowledge of the immune system is integral to today's greater understanding of infectious diseases and cancer.

Hood's drive to develop a quartet of automated instruments in the 1980s, along with Mike Hunkapiller and others, ranks among the greatest contributions to biology of the past century. The inspired work on the automated DNA sequencer made scientists believe that sequencing the entire genome was within reach. Applied Biosystems continuously improved upon the early prototype tools until they were good enough to complete the first human genome.

To his critics, Hood's accomplishments weren't what he made them out to be. And the turmoil he left in his wake was sometimes self-defeating. Like tech visionary Steve Jobs, Hood attracted critics who saw him as a narcissist, a self-promoter, a reckless manager, and a callous, self-centered jerk. The critics see a man, in his later years, prone to hype and sloganeering. His science is no longer first-rate, they say.

There is truth in both interpretations.

Hood's personality undermined him in particular ways. Impatience and an unwillingness to play on a big team, for example, cost him the opportunity to be a leader of the Human Genome Project.

Hood enjoyed being associated with a grand scientific under-taking like the Human Genome Project. He often reminded people of his pioneering role. His automated technologies enabled the genome project, and his advocacy helped persuade Congress to launch it. But when a political appointee called to ask if he wanted to run the federal project after James Watson resigned, Hood's answer was no. It may have been for the best. With Hood's single-minded focus and lack of empathy for what others think and feel, he would have struggled in a political environment like Washington, DC. This was also a fifteen-year project. An impatient man by nature, Hood couldn't possibly stick to the end with something taking that long. He enjoyed diving in at certain moments when there were intellectual puzzles to solve. He generated ideas. He could see how they fit into coherent long-term strategies. He mobilized teams to jump-start the ideas. And then he was on to the next thing. That was one way to stay on the cutting edge. But at the end of the Human Genome Project, the Gairdner Foundation honored eight scientists who made the biggest contributions. Hood wasn't among them. His UW labs completed about 1 percent of the sequencing.

As Mark Davis, a Stanford University immunologist and former Hood graduate student, said:

> The tragedy of Lee, and I would call it that, was that he was perfectly poised to lead the Human Genome Project. He was very much in tune with genes and genetics, and the technology and sequencing. But he kind of walked away.

> He dabbled. He was a dabbler. … He's been a motivator and an idea person, but he gets bored with the nuts and bolts, and the personal aspect of managing.

Hood's next idea hasn't yet lived up to its lofty promises. He moved on to systems biology, one aspect of the new world of biology

that the genome cracked open. But this was dizzyingly complicated. Fifteen years after the founding of the Institute for Systems
Biology, his groups had a long list of peer-reviewed publications,
but scientists still didn't have accurate predictive computer models
for what happens when certain stimulus (like a drug) interacts
with a human cell. The ultimate payoff with personalized medicine
is still far off—at least in terms of large-scale implementation in
health care.

Bill Gates, the most important benefactor of Hood's career,
turned his money and attention toward a new cause of global health.
Other scientists emerged as leaders in the era of genomics, such as
Eric Lander at the Broad Institute in Cambridge, Massachusetts.
Although Hood remained a provocative, strategic thinker in his sev-
enties, his influence faded.

"Nowadays there are all these people looking at gene data and
looking at patterns, but [Hood] didn't get the central role in that
research that he wanted," Gates said in 2014. "He predicted all that
stuff, but he's not the only one."

Hood's teams weren't always first with key technology innovations. But before almost anyone, he foresaw biology becoming data
driven, fast, and team oriented. During his two-decade run on the
Caltech faculty, Hood crafted an ambitious strategy for four automated instruments that would together transform biology.

As he predicted, those instruments became workhorse tools for
the biotechnology industry.

• The protein sequencer was an important tool to speed up development of the first biotech drugs. Hood's work to improve automated protein sequencing enabled Stanley Prusiner to elucidate the
concept of infectious proteins that cause "Mad Cow" disease.

• Automated cell-sorting machines that Hood oversaw now routinely help scientists zero in with enough resolution to find a single
abnormal cell out of a sample with billions.

• Gene expression machines enable researchers to see when an otherwise normal gene expresses its information in a damaging way.

• The DNA sequencer, the most important tool, gave the world a greater understanding of the underlying code of life.

That understanding grows as DNA sequencing gets exponentially faster and cheaper. Ed Lazowska, the Bill & Melinda Gates chairman of computer science and engineering at the University of Washington, stayed close to Hood for more than twenty years, as he thought about how his discipline of computer science could approach tough problems of biology. Lazowska wrote in a 2006 letter to the National Academy of Engineering:

> Four discoveries over the last fifty years have served as milestones in deciphering the digital code of DNA: the discovery of the structure of DNA in 1953 by Watson and Crick, the development of manual sequencing methods in 1975 by Sanger and Gilbert, the development of automated DNA sequencing in 1986 by Hood and his colleagues, and the completion of the Human Genome Project in 2003. The development of automated sequencing made the Human Genome Project possible.

> Before Lee Hood invented the automated DNA sequencer, it took thirty years to sequence the genome of the cold virus. Recently, it took less than a day to sequence the genome of the SARS virus! Before Lee Hood invented the automated DNA sequencer, it cost (by one authoritative estimate) $180 million to sequence the first gene.

> Today, it costs $6 to sequence a gene!

What cost six dollars at the time of that letter would cost pennies a few years later.

The speed of sequencing technology in the first decade of the twenty-first century improved at a rate faster than Moore's law—the observation that the computing power of a chip doubles every eighteen to twenty-four months. Nathan Myhrvold, the chief technology officer of Microsoft and an admirer of Hood, marveled at the change. In a 2012 interview, he said:

> One of my sons is now a graduate student in systems biology at Harvard. They don't even run their own sequencers anymore. A company sends a courier to pick things up. There's a box in the lab for samples. If my son calls before 3:00 p.m., usually by midnight he has sequencing results. He gets them e-mailed. If the results are interesting, he'll go back to the lab and set up a new set of experiments for the next day.
>
> It's just insane how productive it is now. ... Not only was the sequencer important, but it inspired all sorts of other things. For my son's generation of biologists, the idea that you use computing, and machines, and all these other things, is completely second nature.

As genomes get ever faster and cheaper to sequence, it becomes possible to compare large numbers of healthy and diseased people. With that information in hand, scientists are slowly getting a better sense of what makes us human and keeps us well. With complete genomes costing about $1,500 at the time of this writing, scientists can now compare genomes from large numbers of people, and mine all that data to look for more and more meaningful information about what the genes are doing.

Hood's signature contribution to science is clearly the DNA sequencer. His contribution has been compared to the microscope, the printing press, the computer microchip. As physicist Freeman Dyson once said: "New directions in science are launched by new tools much more often than by new concepts. The effect of a concept-driven revolution is to explain old things in new ways. The effect of a tool-driven revolution is to discover new things that have to be explained."

But Hood's legacy is broader still. He changed the way biologists think about their own discipline. By introducing biology to automation and the computer age, he helped scientists see new possibilities.

"People who generate new and pervasively useful instruments that enable you to do things you couldn't do before—that ranks very high in the scientific community," said Daniel Kevles, the Yale science historian. "To enable people to see what they couldn't see before is crucially important."

David Baltimore, the Caltech biologist and Nobel Laureate who has known Hood for more than forty years, summarized it this way:

> He leaves a legacy of having thought properly about DNA rearrangement, having seen it, having used technology to probe the molecular basis of immunology. He could well have [shared in the Nobel Prize].
>
> Then there was the incredible job he did in bringing automation and technical prowess to biology. He did it when other biologists were kicking and screaming. Today, nobody doubts that you need sequencers and mass spectrometers and all these things Lee either originated or was one of the very first to use. He's been a sort of Pied Piper bringing everybody behind him for a good part of forty years.

Now everybody's talking about systems biology.
There was nobody talking about systems biology
before Lee.

With systems biology, Hood sought to understand biology in a broader context, to figure out how networks of genes work together in symphony to create a whole organ, a whole organism. It was like imagining how a set of mechanical parts come together to make a finely tuned automobile. In the industry, this "provided an appropriate counter to the usual reductionist tendencies in science," said Francis Collins, the director of the National Institutes of Health.

Through systems biology, Hood drove an approach that some call "discovery science." In certain ways, this was a different spin on the scientific method itself. It wouldn't be efficient to form a hypothesis, or educated guess, and then run an experiment to gather data that tests the hypothesis. Instead, researchers would first gather enormous volumes of data, then use computers to sift through it to see whether there might be an interesting pattern to test with further experiments or data analysis. "Lee has always been interested in the power of very large data sets. It's an enduring theme in his research," Roger Perlmutter said. "Time and again, he focused on technology to create large data sets to ask bigger questions."

The bigger questions Hood was after? They focused on what UW's Maynard Olson called "informational" rather than "mechanistic." Olson described the distinction this way:

A mechanistic question is "What part of the antibody molecule interacts with the antigen?" Or "How does that trigger downstream events in cellular immune response?" The informational question is "We're confronted with a vast array of antigens. Where does the information come from that we use to build molecules that will bind strongly to the antigens?"

Both are important. But as mechanisms increasingly get worked out, more attention has turned to the information embedded in the DNA, RNA, and proteins.

Late in Hood's life, some thought of him as a technologist first and a biologist second. That was backward. Hood always thought first about hard questions in biology and then imagined technologies that could help get the answers and unveil new possibilities.

Was Hood one of the most influential biologists of the past century? "Without a doubt," Baltimore said.

Hood's impact went far beyond the lab.

He coauthored textbooks that were accessible to undergraduates and graduate students. Occasionally, students would write to him decades later saying the textbooks inspired them to pursue careers in biology. He gave stirring scientific talks. He wasn't afraid to step out of his comfort zone, crossing disciplines, enlisting smart people of all stripes to help tackle big problems in biology. He helped start and advise dozens of companies.

Not only did Hood's work on the DNA sequencer catalyze important science, it created a lot of business opportunities. Caltech alone made more than $100 million in royalties off his lab's inventions. That cash enabled Caltech to establish an aggressive technology transfer office that had the wherewithal to seek patents for more than one hundred companies that might otherwise never have been started.

Business, to Hood, was an instrument to extend science into society. He wasn't among the scientists content to communicate their work within the confines of the academy, to advance knowledge among a handful of peers who understand it. "I believe every scientist has an obligation to transfer their knowledge to society," Hood once said. "Creating companies (and licensing intellectual property) is one effective way to do so."

Making the connection between science and society led Hood to engage in K–12 education, long-running conversations with ethicists, and countless media interviews. He was almost evangelical in his pro-science and pro-technology advocacy, but he appreciated thoughtful dialogue from people in the humanities. When science opened up new ethical questions, he welcomed the input of science historians and ethicists. Some scientists could be arrogantly dismissive of ethicists looking over their shoulders, but Hood never claimed that scientists had all the answers. At the dawn of the genomics age, while accepting the 2002 Kyoto Prize, Hood said:

> Predictive and preventive medicine will pose striking social, ethical, and legal issues for society. How will society treat seventy- to ninety-year-olds who are still vital, productive, and creative? How will medical schools train physicians who in fifteen to twenty-five years will be practicing predictive and preventive medicine? … How will we deal with issues of genetic privacy?

> The engineer will soon be able to engineer himself or herself through germ-line genetic engineering. The germ line is the DNA passed on to future generations. Hence, engineering of the fertilized human egg, for example, means that the germ-line genetic changes will become a permanent part of human heredity. Is it appropriate to use germ-line genetic engineering to avoid disease or to improve the human condition (e.g., increase intelligence)?

> This debate will be a major societal issue for the future.
> ... The most reasonable approach to dealing with most
> of these issues is to have a thoughtful, informed, and
> rational public.

Hood devoted more than fifty years of his life to science education for young people. He wanted to share his enthusiasm and cultivate subsequent generations of scientists. But he also wanted to do his part to create a more science-literate society that could intelligently sort through the proper uses of science and technology. He generously gave his time to middle school and high school students. He influenced education officials to make science education less about memorizing facts from textbooks and more about asking imaginative questions. When he started his own research institute, even when budgets were tight, he remained committed to the cause. His science education efforts, often with budgets of more than $1 million a year, concentrated on providing schools with inquiry-based curricula, materials, and continuing education opportunities for more than a thousand K–12 science teachers in Washington State. They spread their learnings among tens of thousands of students. He loved it. "I have had the enormous pleasure of seeing kids become infected with the spirit of inquiry-based thinking," Hood once said. "K–12 science education is a forever vocation."

Hood was deeply dismayed to see America drift toward anti-intellectualism in his later years. While science and humanities continued to veer onto divergent paths, Hood insisted on maintaining relationships and open dialogue with people on both sides of that great divide. This is one part of Hood's career that's impossible to quantify, but real nonetheless. At a time when many biologists didn't care to or didn't bother to explain their work to lay audiences, Hood thrived on making biology accessible to everyone. He was one of biology's greatest communicators.

Hood wasn't a creative genius who hurled theoretical thunderbolts, like Albert Einstein. While there is no standard definition for genius, many scientists reserve the compliment for those rare individuals who make unprecedented leaps of insight that otherwise might never have occurred. Hood was brilliant and driven, but his impact was of a different sort. Often, he was ahead of the curve, working on problems that others would have solved sooner or later. What Hood did exceptionally well was assemble teams, raise money, and inspire scientists to carry out plans that looked further down the road. "He is a big thinker," said NIH Director Collins, "and he's always a few steps ahead of the field." Hood had superb instincts for where things were headed.

He created a lab environment where great things could happen. He left details to others, and he expected his students to figure things out for themselves. That trust and faith in other people could create comical breakdowns in communication and general chaos. But it also stimulated many to tap an inner resourcefulness they never knew they had. His enthusiasm was infectious. Science was a joy to him. He motivated many students to do the best work of their careers.

At the same time, Hood made many other students and peers uncomfortable. His self-centered behavior and grabs for credit were demoralizing at worst, irritating at best. When I met James Watson, the codiscoverer of the DNA helix, at a Seattle event organized by billionaire Paul Allen, the provocative elder biologist scoffed at the idea of meeting with Hood. "Why would I want to listen to him talk about himself?" Watson said.

The whirlwind nature of Hood's whole life, working nonstop with manic intensity, was exhausting to even the hardiest graduate students. His style challenged popular notions of what scientists do. Where some saw a dynamo breaking open new fields, others saw a dabbler, an opportunist—someone who skimmed the surface with knowledge a mile wide and an inch deep.

Ron Davis, a distinguished Stanford biologist and technologist, described how inconsistent with science's behavioral norms Hood's style was:

> You go to every meeting. You write books, write review articles. You stay in a certain territory. …
> Many scientists are associated with a signature achievement. I don't think that happened for Hood, as it has for other scientists. It's in part because he didn't stay with [things]. He developed [the automated DNA sequencer]. But then he turned it over to ABI. … He would move on to the next thing when it was clear it was going to work, but often before other people recognize it's going to work.
>
> A lot of scientists are very critical of that. Their mind-set is of this signature accomplishment—that you stay with things. It's like professionalism. If you're a scientist, you have to pick an area and dominate. In their minds, he was not being a scientist. … Lee Hood's talent is to lay new groundwork in new areas. That's what he's done his entire career.

Put another way? "Lee starts things. We have a shortage of people in science that start things," said Maynard Olson. "We have unbelievable human resources in American science to do things once they are going, but it's hard to start new things." Those who dwell on Hood's inability to finish things miss the point, Olson said. Other people are good at doing that.

Hood was at his best when he had these other people—technically capable, disciplined doers—by his side. His greatest success came as a younger scientist when he found and retained outstanding operating deputies, like Mike Hunkapiller. Without an exceptional deputy getting things done, Hood could seem adrift. It was a common complaint in his later years.

"A lot of people, not just his enemies, were exasperated with him," Baltimore said. "It's not so much self-promotion as 'Where's the beef?'" Hood had a tendency, Baltimore said, to stand up at scientific meetings and promise that some new experimental data, or big advance, was just around the corner. Months or years later, when colleagues asked what happened, there was nothing to say. "It never comes out. You never see it. Because he *wasn't* almost there," Baltimore said. "Perhaps it was unjustified optimism."

He wasn't almost there with personalized medicine, either—not at large scale. By 2016, sequencing still hadn't made its way into mainstream medicine. Significant technological work still remained to make it good enough to help predict, prevent, and personalize treatment for people with cancer or other common diseases.

However, a few significant strides have been made:

• Genentech's Herceptin for people with a genetic variation that causes breast cancer, Novartis's Gleevec for chronic myeloid leukemia, and Pfizer's Xalkori for a type of lung cancer were significant advances.

• Vertex Pharmaceuticals offered the first treatment for one of the underlying genetic defects in cystic fibrosis—a fatal lung disease. The drug was remarkably effective in helping patients breathe better, maintain healthy weight, and otherwise remain active young adults.

• By the mid-2010s, a genomic test was introduced to scan for more than three hundred genetic variations that are known to drive cancer growth. The test was developed by Foundation Medicine, a company based in Cambridge, Massachusetts, that was backed by Gates and other high-profile investors. Hollywood actress Angelina Jolie took a test from another company in 2013 that revealed she carried a genetic mutation increasing her risk of breast cancer. Jolie had a preventive double mastectomy—stirring a national debate and prompting many women to get their own genetic tests. It was quite a leap in the public debate. Just two years earlier, when Apple cofounder Steve Jobs paid $100,000 to sequence his genome, he was looking for genetic variants that might be driving his pancreatic cancer. His decision at that time was considered by many to be a novelty.

• Genetic tests that look at the risk of breast cancer recurrence became widely available in the second decade of the twenty-first century. Genomic Health's Oncotype Dx and Agendia's MammaPrint could evaluate a set of genes that offered high predictive value of a recurrence. Using that data, doctors and patients could also see when a round of preventive chemotherapy was likely to succeed or not. Many women were able to avoid toxic, costly, and unnecessary chemotherapy based on the results.

• One Stanford researcher, in an N-of-1 study of himself, found in 2012 that he was tilting toward a diabetic state, after sequencing his genome and looking at a battery of RNA and protein tests. His doctor didn't see any classic symptoms, so the prediction seemed absurd. Sure enough, prediabetes symptoms became apparent within a few months. The researcher was able to switch back to a healthy state by improving his diet and exercise habits. It was an early test case for the power of predictive, preventive medicine.

Genomics certainly has been overhyped at times and has had its disappointments. Google-backed 23andMe persuaded hundreds of thousands of people to have parts of their genomes scanned for as little as ninety-nine dollars. That's when it ran afoul of regulators, who accused it of going too far in providing information of shaky, uncertain medical value. After two years of work with the Food and Drug Administration on how to responsibly communicate genomic information to the public, 23andMe reemerged in

2015 with a slimmed-down test of genes that speak to a person's carrier status—genes that could pass down a serious disease to future generations.

Twenty years ago, Hood predicted there would be more progress for the personalized medicine movement. Society is still wrestling with thorny issues of cost, access to care, privacy. Many scientists and physicians worry about genetic information being misinterpreted or overhyped by businesses selling to less educated consumers.

Plenty of technical hurdles remain with interpreting, visualizing, and applying genomic data to health care. The legal, business, and political barriers are formidable. Scientists will need to generate mountains of evidence to tear down some of those walls. Even when they do, turning molecular understanding into clinical treatment is not a straightforward process.

"Proteins often behave differently in test tubes than in cells, and cells behave differently in culture than as part of a vital organism," two physician-scientists wrote in a 2000 commentary on the dawn of genomic medicine. "Finally, a patient's experience of disease reflects more than simply an underlying biological defect."

The immense challenge is that diseases are amazingly complex, involving the interplay of multiple genes and the environment—and these can change dramatically over time. Genes don't dictate a person's health destiny. Behavior and environment have a lot to do with it, as anyone familiar with classic nature versus nurture debates knows. Rarely can a scientist or physician point to a single aberrant gene and then prescribe a drug that works on that single gene like a magic bullet.

Science will lurch forward in fits and starts, as it always has. Physicians will be on the front lines of interpreting this massive accumulation of data and deciding how to best act upon it. Only over time and with repeated experimentation will genomics data be transformed into reliable working medical knowledge.

Much of the work of systems biology—teasing apart the connections and how they change over time—remains to be done.

Nevertheless, the proponents of genomic-based personalized medicine have reasons to be optimistic. The new biology that Hood imagined and championed is here. The payoff may not have been fully realized in Hood's lifetime. But it will arrive. Hood's technology-driven, data-intensive approach to biology is alive in the hands of tens of thousands of scientists around the world. Their ambition and their capacity for improving human health become more expansive each day.

Acknowledgments

This book depended heavily on a team with diverse talents.

For starters, there's my wife, Tracy Cutchlow. She's an accomplished book editor, having made significant contributions to the best-selling neuroscience books *Brain Rules* and *Brain Rules for Baby*, as well as *Modernist Cuisine*, on the science of cooking. She was my most important collaborator from start to finish. Her expertise is truly multidimensional. She acted as an almost daily sounding board for the past four years. She performed the crucial developmental editing job, making many organizational improvements.

Robert L. Simison provided essential editing and guidance, particularly on translating the science and technology to a lay audience. Simison—a veteran of the *Wall Street Journal* and my former editor at Bloomberg News—is first-rate.

Seattle photographer Brian Smale captured the gripping cover image. Todd Bennings provided the graphic design vision necessary to bring it to life on the cover. George Church graciously wrote the foreword. Carrie Wicks handled the copyediting with ease.

Several archivists were helpful, particularly Loma Karklins at Caltech and Gary Lundell at the University of Washington. David Turim helped with finding legal documents. The Seattle Public Library's free resources were invaluable.

At the Institute for Systems Biology, Sissy Bouchard, Hsiao-Ching Chou, and Jeremy Johnson all provided important assistance in tracking down documents and photos.

Several independent readers provided critical feedback on early drafts of the manuscript, including Kim Emmons, Christopher Henney, Stewart Lyman, Kyle Serikawa, Todd Smith, and Sarah Warren.

Members of the Hood family were helpful. Hood's sister, Doral Vial, and his brother Myron, in particular, provided a treasure trove of family photos and letters. Hood's children, Eran and Marqui, graciously agreed to be interviewed.

Many others provided important words of support, especially in the early going. Helpful advice and encouragement came from Melissa Allison, Deborah Blum, Todd Burklin, David Ewing Duncan, Wier Harman, Matthew Herper, Sakina Hussein, Thong Le, Victor McElheny, Mark Pearson, Tom Paulson, Peter Rinearson, Amina Al-Sadi, Marcie Sillman, Mike Stroh, and Theral Timpson.

More than a hundred people agreed to be interviewed for this book, too many to list here. They were all generous with their time and insights. During especially busy times, when I wondered whether it made sense to finish the project, I felt I owed it to them to keep going. Thanks to all who made this such a fulfilling undertaking.

Endnotes

―――――

CHAPTER 1. MR. GATES, I QUIT

major coup: Bishop, Jerry E. "William Gates's Gift Spurs Shift by Star Biologist." Wall Street Journal. 8 Oct. 1991. Page B1

accelerated the whole field and *creative anarchy:* King, Ralph T. Jr. "Gene Machines: An Eclectic Scientist Gives Biotechnology a Fast Assembly Line." Wall Street Journal. 30 May 1995 Page A1

how are you going to fund?: Interview with Lee Hood at his Seattle condominium. 20 Oct. 2012

Gates said later: "I definitely disappointed Lee" Interview with Bill Gates, responses in writing, 25 Oct. 2014

Bill had not, at that time: Interview with Roger Perlmutter at the Institute for Systems Biology in Seattle. 24 Jan. 2013

I always joke: Hood, Eran. Video interview shown at Hood's 70th birthday party, W Hotel, Seattle. 10 Oct. 2008

CHAPTER 2. GROWING UP IN BIG SKY COUNTRY

Montana is a wonderful place: Johnson, Peter. "Shelby High Grad a Shining Star of Science." Great Falls Tribune. 29 Sept. 1995. Page A1

It is a land of the big sky: Harper, George. "Western Montana Lives. George Harper Loved State from the Start." The Missoulian. 16 May 2011. http://missoulian.com/lifestyles/hometowns/western-montana-lives-george-harper-loved-state-from-start/article_1696af14-7fd8-11e0-918e-001cc4c03286.html

Steinbeck, John. "Travels with Charley: In Search of America." Penguin Books. Paperback version. 1980. http://www.amazon.com/Travels-Charley-Search-America-Steinbeck/dp/0140053204

What Steinbeck Saw: Tyer, Brad. "Opportunity, Montana: Big Copper, Big Water, and the Burial of an American Landscape." Beacon Press. 2013. http://www.amazon.com/Opportunity-Montana-Copper-American-Landscape/dp/0807003298

no idea of tradition: Interview with Irving Weissman at Accelerator Corp. in Seattle. 1 Apr. 2013

cut a handsome figure: "Roy Wadsworth: One Eye Open for Mischief." Carbon County News (Red Lodge, MT) 1966.

he claimed to have met the mountain man: Interview with Doral Vial and Myron Hood at Vial's home in Gig Harbor, WA. 28 Nov. 2012

Leroy is really a grand baby: Hood, Myrtle. Letter to Roy Wadsworth and Florence Freeman from Missoula, MT. 12 Oct. 1938

paid the teacher from their own pockets: "Former Resident of Elso Area Dies at Fort Harrison." The Roundup Record-Tribune (Roundup, MT) 30 Apr. 1959

perhaps in part due to my brother Glen: Hood, Leroy E. "My Life and Adventures Integrating Biology and Technology." A Commemorative Lecture for the 2002 Kyoto Prize in Advanced Technologies. 10 Nov. 2002. https://www.systemsbiology.org/download/2002Kyoto.pdf

CHAPTER 3. THE GOLDEN BOY OF SHELBY

they lured boxing champion: Marcus, Norman. "Dempsey vs. Gibbons: The Fight that Ruined a Town." 21 Nov. 2012. Boxing.com. http://www.boxing.com/dempsey_vs._gibbons_the_fight_that_ruined_a_town.html

classmates voted him best all-around: Shelby High School Yearbook, Coyote '56. Spring 1956

honor roll: Shelby Promoter. 1 Mar. 1956

sometimes exaggerated stories about his football glory days: Dietrich, Bill. "Future Perfect -- Thanks To Bill Gates' $12-Million Endowment, Scientist Leroy Hood Continues His Search For A New Genetic Destiny." The Seattle Times. Pacific Magazine. 9 Feb. 1992. Page 6. http://community.seattletimes.nwsource.com/arc hive/?date=19920209&slug=1474735

average score of 40-6: Football. Shelby High School Yearbook, Coyote '55. Spring 1955

I was good, but not great: Johnson, Peter. "Shelby High Grad a Shining Star of Science." Great Falls Tribune. 29 Sept. 1995. Page A1

he could read the defense: Interview with Arlene (Lingle) Littlejohn, Beth (Thompson) Aklestad, and Dwayne Irvin in Shelby, Montana. 6 Sept. 2013

Olson intervened. He came over to visit Hood's parents: Hood, Leroy E. Interviewed by David A. Valone. Pasadena, California. Aug. 1 and Aug. 3, 1994; Jan. 13 and Mar. 5, 1995. Oral History Project, California Institute of Technology Archives. Unpublished copy provided by Leroy Hood.

among the forty finalists: Leroy Hood is Science Finalist." Shelby Promoter. 2 Feb. 1956. Page 1.

the high school band played: Ciotti, Paul. "Fighting Disease on the Molecular Front. Leroy Hood Built A Better Gene Machine And The World Beat A Path To His Lab" Los Angeles Times Magazine. 20 Oct. 1985. Page 18. http://articles. latimes.com/1985-10-20/magazine/tm-14242_1_leroy-hood-s-molecular-biology

a scholarship from the General Motors Foundation: Hood, Leroy E. Interviewed by Arnold Thackray, David C. Brock, and Arthur Daemmrich. New Orleans, Louisiana. 19 Mar. 2002. Beckman Heritage Project, Chemical Heritage Foundation.

CHAPTER 4. WELCOME TO THE BIG TIME

symptoms of smog complex: History of Smog. LA Weekly (Los Angeles, CA). 9 Sept. 2005 http://www.laweekly.com/2005-09-22/news/history-of-smog/

took less than 20 years to rise to prominence: Goodstein, Judith R. "Millikan's School. A History of the California Institute of Technology." W.W. Norton & Company. 1991

all-around patron saint: Goodstein, David. "On Fact and Fraud: Cautionary Tales from the Front Lines of Science." Princeton University Press. 2010

the reason most people treasure: Interview with Mark M. Davis at Stanford University. 28 Jan. 2013

not emotionally appealing: Dodge, Tom. "Purposes of Frosh Rotation

Reemphasized by Tech Chief." The California Tech. 28 Sept. 1956. Page 1.

orientation camp: "Faculty, Students Tell of Frosh Life at Caltech." The California Tech. 28 Sept. 1956

a formal program: "Frosh to Meet Local Belles." The California Tech. 28 Sept. 1956. Page 1.

one of the strongest freshman classes that Tech has yet had: "The Big T." Caltech Yearbook. (Pasadena, CA) 1957

willing to try new things: Interview with Eric Adelberger by phone. 8 Apr. 2014

sleep is usually the first thing to go: Yale Daily News Staff. "The Insider's Guide to The Colleges: Students on Campus Tell You What You Really Want to Know." 32nd edition. St. Martin's Griffin. 2006. http://books.google.com/books?id=wBm_zuf-ce8C&pg=PA1&lpg=PA1&dq=The+Insider%E2%80%99s+Guide+to+The+Colleges:+Students+on+Campus+Tell+You+What+You+Really+Want+to+Know&source=bl&ots=lwZDCc3hJK&sig=EcH65s114IkVq8m3O8DE_k_wD6E&hl=en&sa=X&ei=K2n_U_jBH8q5ogSdvIHYBQ&ved=0CIkBEOgBMAg#v=onepage&q=caltech&f=false

He really had a leg up: Interview with Daniel Kevles by telephone. 3 Nov. 2014

hunger to learn: Interview with Irving Weissman at Accelerator Corp. in Seattle. 1 Apr. 2013

strange things have happened to this brother of yours: Hood, Leroy E. Personal letter to Doral Vial. 10 Sept. 1959

wonderful feeling of knowing you have given to this world: Hood, Leroy E. Personal letter to Doral Vial. 9 Oct. 1960

after Hood graduated: Feynman, Richard Phillips. Complete Dictionary of Scientific Biography. 2008. Encyclopedia.com. (Retrieved 25 July 2014). http://www.encyclopedia.com/topic/Richard_Phillips_Feynman.aspx

pretty terrifying to undergraduates: Hood, Leroy E. Interviewed by David A. Valone. Pasadena, California. Aug. 1 and Aug. 3, 1994; Jan. 13 and Mar. 5, 1995. Oral History Project, California Institute of Technology Archives. Unpublished copy provided by Leroy Hood.

mercilessly caustic: McElheny, Victor K. "Watson and DNA: Making a Scientific Revolution." Perseus Publishing. 2003.

don't do fashionable science: Segre, Gino. "Ordinary Geniuses: How Two Mavericks Shaped Modern Science." Viking Adult. 2011

always had wonderful insights: Cold Spring Harbor Oral History Collection. CSHL Digital Archives. Interview with Leroy Hood. 1 Jun. 2003 http://library.cshl.edu/oralhistory/speaker/leroy-hood/

appreciate the marvelous beauty and complexity: Hood, Lee. "A Personal Journey of Discovery: Developing Technology and Changing Biology." Annual Review of Analytical Chemistry. Annual Reviews. (Palo Alto, CA). 20 Feb. 2008 http://www.ncbi.nlm.nih.gov/pubmed/20636073

the fatherless young man finds: Rhodes, Richard. "The Making of the Atomic Bomb." Simon & Schuster Paperbacks. 1986. Page 143. http://www.amazon.com/The-Making-Atomic-Bomb-Anniversary/dp/1451677618

in this nascent discipline: Judson, Horace Freeland. "The Eighth Day of Creation: The Makers of the Revolution in Biology." Simon & Schuster. 1979

Hinrichs Memorial Award: Caltech website. Description of Prizes and Awards. Retrieved 1 Dec. 2014 https://deans.caltech.edu/Services/LeadershipPrizes

CHAPTER 5. THE MAKING OF A YOUNG BIOLOGIST

famous physician-scientist: Ballard Jr., Frederic. "William Barry Wood." Harvard Crimson. 8 Jan. 1963 http://www.thecrimson.com/article/1963/1/8/william-barry-wood-pwilliam-barry-wood/

a couple of competing hypotheses: Hendricks, Melissa. "The Big Biology of Leroy Hood." Johns Hopkins Magazine. Feb. 1990

what's wrong with medical school education: Hood, Leroy E. Interviewed by David A. Valone. Pasadena, California. Aug. 1 and Aug. 3, 1994; Jan. 13 and Mar. 5, 1995. Oral History Project, California Institute of Technology Archives. Unpublished copy provided by Leroy Hood.

despite a few Cs: Johns Hopkins University transcript. Received 14 Apr. 2014

the cheapest Chinese restaurant" Interview with Steve Rosenberg by phone. 19 Mar. 2013

Dreyer invited Hood to be his first graduate student: Dreyer, William. Interview by Shirley K. Cohen. Pasadena, California, February 18–March 2, 1999. Oral History Project, California Institute of Technology

Archives. Retrieved 28 July 2014 from the World Wide Web: http://resolver.caltech.edu/CaltechOH:OH_Dreyer_W

married in a small ceremony: "Rite Unites Miss Logan, Leroy Hood." Great Falls Tribune. 2 Feb. 1964

North Holliston Avenue: Property records http://publicrecords.directory/profiles/valerie-hood.1735707.html

a whole night lab culture: Hendricks, Melissa. "The Big Biology of Leroy Hood." Johns Hopkins Magazine. Feb. 1990

kind of a klutz: Dreyer, William. Interview by Shirley K. Cohen. Pasadena, California, February 18–March 2, 1999. Oral History Project, California Institute of Technology

Archives. Retrieved 28 July 2014 from the World Wide Web: http://resolver. caltech.edu/CaltechOH:OH_Dreyer_W

hampered in his grand plans: Hendricks, Melissa. "The Big Biology of Leroy Hood." Johns Hopkins Magazine. Feb. 1990

affectionately known to colleagues: Berg, Paul and Singer, Maxine. "George Beadle: An Uncommon Farmer. The Emergence of Genetics in the 20th Century." Cold Spring Harbor Laboratory Press. 2003. http://www.amazon.com/ George-Beadle-Uncommon-Farmer-Emergence/dp/0879697636

a provocative paper: Dreyer, William J.; Bennett, J. Claude. "The Molecular Basis of Antibody Formation: A Paradox." Proceedings of the National Academy of Sciences. (Washington, D.C.) 15 July 1965. http://www.jstor.org/discover/10.2307 /72691?uid=3739960&uid=2&uid=4&uid=3739256&sid=21104533621663

borrow a quote from one of Abraham Lincoln's addresses: Dreyer William J., Gray WR, Hood LE. "The Genetic, Molecular, and Cellular Basis of Antibody Formation: Some Facts and a Unifying Hypothesis." Cold Spring Harbor Symposia on Quantitative Biology. 1967. 32:353-367. http://symposium.cshlp. org/content/32/353

if you want a protégé to take off: Dreyer, William. Interview by Shirley K. Cohen. Pasadena, California, February 18–March 2, 1999. Oral History Project, California Institute of Technology

Archives. Retrieved 28 July 2014 from the World Wide Web: http://resolver. caltech.edu/CaltechOH:OH_Dreyer_W

Grand Canyon country: Hood, Leroy E. "Immunoglobulins: Structure, Genetics, and Evolution." PhD thesis. California Institute of Technology. Submitted 29 Nov. 1967. PhD awarded 1968. http://thesis.library.caltech.edu/3561/1/Hood_ le_1968.pdf

yellow berets: Interview with Gilbert Omenn at the Institute for Systems Biology in Seattle. 12 Aug. 2013.

Additional reading:

Hood, Leroy E. "My Life and Adventures Integrating Biology and Technology." A Commemorative Lecture for the 2002 Kyoto Prize in Advanced Technologies. 10 Nov. 2002. https://www.systemsbiology.org/download/2002Kyoto.pdf

Kuhn, Thomas S. "The Structure of Scientific Revolutions." 1962. University of Chicago Press. http://en.wikipedia.org/wiki/ The_Structure_of_Scientific_Revolutions

CHAPTER 6. A RISING STAR FOR THE BIOTECH ERA

Salary conversions, adjusted for inflation: U.S. Department of Labor, Bureau of Labor Statistics. Consumer Price Index Inflation Calculator. Retrieved 29 July 2014. http://www.bls.gov/data/inflation_calculator.htm

Beadle resisted developing tunnel vision: Berg, Paul and Singer, Maxine. "George Beadle: An Uncommon Farmer. The Emergence of Genetics in the 20th Century." Cold Spring Harbor Laboratory Press. 2003. http://www.amazon.com/George-Beadle-Uncommon-Farmer-Emergence/dp/0879697636

hiring was noted: New Faculty. "A Report for the Year 1969-1970 on the Research and Other Activities of the Division of Biology at the California Institute of Technology." Page 1.

almost tripled during the 1960s: History of Congressional Appropriations. National Institutes of Health. 1960-1969. Retrieved 29 July 2014 http://officeof-budget.od.nih.gov/pdfs/FY08/FY08%20COMPLETED/appic3806%20-%20 transposed%20%2060%20-%2069.pdf

Congress tripled it again: History of Congressional Appropriations. National Institutes of Health. 1970-1979. Retrieved 29 July 2014 http://officeofbudget. od.nih.gov/pdfs/FY08/FY08%20COMPLETED/appic3806%20-%20trans-posed%20%2070%20-%2079.pdf

By 2004, it was forty-two: Average Age of New Investigators at Initial R01 Equivalent Award. National Institutes of Health. Data retrieved 29 July 2014. https://www.google.com/webhp?sourceid=chrome-instant&rlz=1C1LENP_enUS512US512&ion=1&espv=2&ie=UTF-8#q=average%20age%20of%20 first%20nih%20grant%201970

first published in 1974: Wood, William B.; Wilson, John H.; Benbow, Robert M.; Hood, Leroy E.; "Biochemistry: A Problems Approach." The Benjamin/Cummings Publishing Company. 1974.

I hear and I forget: Wood, William B. "Genesis of Biochemistry: A Problems Approach." Cell Biology Education, American Society for Cell Biology. Spring/Summer 2002. http://www.ncbi.nlm.nih.gov/pmc/articles/PMC118368/

Neither of us was interested in single genes: Interview with Eric Davidson at his Caltech office. 31 Jan. 2013.

it used a novel spinning cup: Bryan Smith, John. "Protein Sequencing by Edman Degradation." Encyclopedia of Life Sciences. 2001. http://www.ehu.es/biofisica/juanma/papers/EdmanDegradation.pdf

I see it in my head: Dreyer, William. Interview by Shirley K. Cohen. Pasadena, California, February 18–March 2, 1999. Oral History Project, California Institute of Technology Archives. Retrieved Feb. 27, 2016 from the World Wide Web:

http://resolver.caltech.edu/CaltechOH:OH_Dreyer_W

God only knows: Jenkins, Philip. "Decade of Nightmares: The End of the Sixties and the Making of Eighties America." Oxford University Press. 2006 http://books.google.com/books?id=Tp9WODvGNScC&pg=RA1-PT19&lpg=RA1-PT19 &dq=god+crawled+laboratory+cambridge+mayor+1976&source=bl&ots=9LP DFdsUII&sig=Vt5bwZHR3m9sqOfWLChVjr2IPZo&hl=en&sa=X&ei=3uR8VN_ 9JIKzogSzh4DoDQ&ved=0CCYQ6AEwAg#v=onepage&q=god%20crawled%20 laboratory%20cambridge%20mayor%201976&f=false

he gabbed: Cooke, Bob. "Antibody Study Reveals Immunity Clues." Pasadena Star-News. 2 May 1971. Page 1 http://newspaperarchive.com/us/california/pasadena/ star-news/1971/05-02/

increasingly obsessed: Interview with Eric Adelberger by telephone. 8 Apr. 2014

enlisted as many as 20: A Report for the Year 1974-1975 on the Research and Other Activities of the Division of Biology at the California Institute of Technology.

incredibly energetic and enthusiastic: Interview with Mike Hunkapiller at Pacific Biosciences in Menlo Park, CA. 15 Oct. 2012

actually a pretty good listener: Interview with Mitchell Kronenberg by telephone. 6 June 2014.

a mentor should provide: Hood, Leroy E. "My Life and Adventures Integrating Biology and Technology." A Commemorative Lecture for the 2002 Kyoto Prize in Advanced Technologies. 10 Nov. 2002. https://www.systemsbiology.org/ download/2002Kyoto.pdf

read the newspaper sports pages: Interview with Eran Hood in Seattle. 29 Nov. 2012

expected to run around the neighborhood: Interview with Marqui Hood by telephone. 20 Feb. 2013

one meal wasn't pot-luck style: Interview with Tim Hunkapiller, Mercer Island, WA. 14 Nov. 2012

CHAPTER 7. THE FIRST BIG MACHINE

Edman and Begg's chemistry work: Edman, Pehr and Begg, Gregory. "A Protein Sequenator." European Journal of Biochemistry. 1967. http://onlinelibrary.wiley. com/doi/10.1111/j.1432-1033.1967.tb00047.x/pdf

Dreyer secured a patent: Dreyer, William J. "Peptide or protein sequencing method and apparatus" U.S. Patent 4,065,412. Issued December 27, 1977 http:// patents.com/us-4065412.html

a gas-liquid phase chamber: Sharrer, Terry. "The Dreyer Peptide and Protein Sequencer." The Scientist. July 2007. http://www.the-scientist.com/?articles. view/articleNo/25189/title/The-Dreyer-Peptide-and-Protein-Sequencer/

An important new phase of protein chemistry is underway: Hunkapiller, Michael W. and Hood, Leroy E. "Direct microsequence analysis of polypeptides using an improved sequenator, a nonprotein carrier (Polybrene), and high pressure liquid chromatography." Biochemistry. May 1978. http://pubs.acs.org/doi/abs/10.1021/bi00604a016?journalCode=bichaw

patent application was whipped together: Hood, Leroy; Hunkapiller, Michael W. U.S. Patent #4,252,769. Filed Dec. 26, 1979, Issued Feb. 24, 1981 "Apparatus for the performance of chemical processes." http://patft.uspto.gov/netacgi/nphParser?Sect1=PTO2&Sect2=HITOFF&p=1&u=%2Fnetahtml%2FPTO%2Fsearch bool.html&r=38&f=G&l=50&co1=AND&d=PTXT&s1=4,252,769&OS=4,252,769&RS=4,252,769

excitement was palpable: Hunkapiller Michael W., Hood, Leroy. "New Protein Sequenator with Increased Sensitivity." 1 Feb. 1980. *Science* 207:523-525 http://www.ncbi.nlm.nih.gov/pubmed/7352258 http://www.sciencemag.org/content/207/4430/523

a blockbuster series of three more papers: Hunkapiller Michael W., Hood, Leroy. "Human Fibroblast Interferon: Amino Acid Analysis and Amino Terminal Amino Acid Sequence" 1 Feb. 1980. *Science*

207:525-526. http://www.sciencemag.org/content/207/4430/525.short

a blockbuster series of three more papers: Zoon KC, Smith ME, Bridgen PJ, Anfinsen CB, Hunkapiller MW, Hood LE. (1980) Amino Terminal Sequence of the Major Component of Human Lymphoblastoid

Interferon. *Science* 1 Feb. 1980 207:527-528. http://www.ncbi.nlm.nih.gov/pubmed/7352260

a blockbuster series of three more papers: Taira HR, Broeze J, Jayaram BM, Lengyel P, Hunkapiller MW, Hood L. Mouse

Interferons: Amino Terminal Amino Acid Sequences of Various Species. *Science* 1 Feb. 1980 207:528-530. http://www.ncbi.nlm.nih.gov/pubmed/7352261

a definitive paper: Hewick RM, Hunkapiller MW, Hood LE, Dreyer WJ. "A gas-liquid solid phase peptide and protein sequenator." Journal of Biological Chemistry. 10 Aug. 1981 http://www.ncbi.nlm.nih.gov/pubmed/7263636

Dreyer was the senior author: Interview with Lee Hood 19 Mar. 2016 (Confirmation that Dreyer deserved to be senior author on key 1981 publication in the Journal of Biological Chemistry describing the protein sequencer's gas-liquid solid phase technology).

the instrument I invented: Dreyer, William. Interview by Shirley K. Cohen. Pasadena, California, February 18–March 2, 1999. Oral History Project, California Institute of Technology Archives. Retrieved Feb. 27, 2016 from the World Wide Web: http://resolver.caltech.edu/CaltechOH:OH_Dreyer_W

we're going to make picks and shovels for the biotech gold rush: Applied Biosystems Company History. "Twenty Five Years of Advancing Science."

a booster rocket: Dietrich, Bill. "Future Perfect -- Thanks To Bill Gates' $12-Million Endowment, Scientist Leroy Hood Continues His Search For A New Genetic Destiny." The Seattle Times. Pacific Magazine. 9 Feb. 1992. Page 6. http://community.seattletimes.nwsource.com/archive/?date=19920209& slug=1474735

once deemed impossible: Bell, John R., Hunkapiller, Michael W., Hood, Leroy E., Strauss James H. "Amino-Terminal Sequence Analysis of the Structural Proteins of Sindbis Virus." Proceedings of the National Academy of Sciences. June 1978 http://www.ncbi.nlm.nih.gov/pmc/articles/PMC392635/pdf/pnas00018-0188.pdf

mechanisms of life itself: Caltech Formally Welcomes The Braun Laboratories to Campus. Caltech News. Feb. 1983. Page 1. http://caltechcampuspubs.library. caltech.edu/2414/1/1983_02_17_01.pdf

they were more supportive of his dreams: Interview with Leroy Hood: America's most productive biologist on philanthropy's role in science. Philanthropy magazine. 31 Jan. 2013

Additional background reading:

A Report for the Year 1977-1978 on the Research and Other Activities of the Division of Biology at the California Institute of Technology."

Hood, Leroy. "My Life and Adventures Integrating Biology and Technology." A Commemorative Lecture for the 2002 Kyoto Prize in Advanced Technologies. https://www.systemsbiology.org/download/2002Kyoto.pdf

Hendricks, Melissa. "The Big Biology of Leroy Hood." Johns Hopkins Magazine. Feb. 1990

Hood, Lee. "A Personal Journey of Discovery: Developing Technology and Changing Biology." Annual Review of Analytical Chemistry. Annual Reviews. 20 Feb. 2008 http://www.ncbi.nlm.nih.gov/pubmed/20636073

CHAPTER 8. BUILDING AN EMPIRE

a protein that seemed to interfere: Kolata, Gina. "Jean Lindenmann, Who Made Interferon His Life's Work, Is Dead at 90." The New York Times. 22 Jan. 2015. http://www.nytimes.com/2015/01/23/us/

jean-lindenmann-made-interferon-his-lifes-work-is-dead-at-90.html

the IF drug for cancer: Cover of Time Magazine. "The IF Drug for Cancer." 31 Mar. 1980. http://content.time.com/time/covers/0,16641,19800331,00.html

you sprinkled on stockbrokers: McDermott, Terry. "101 Theory Drive: The Discovery of Memory" Vintage Books. 2010. http://books.google.com/books?id=jExDDLcgazYC&pg=PA106&lpg=PA106&dq=interferon+101+theory+drive+money&source=bl&ots=1EhRLF55e4&sig=Dj2p1Y74GguL1eBX1EiuZDY22_4&hl=en&sa=X&ei=4KOAVNOMKKPpiQKXm0CwAQ&ved=0CB4Q6AEwAA#v=onepage&q=interferon%20101%20theory%20drive%20money&f=false

perhaps the most expensive substance in the world: Meredith, Dennis. "Cracking the Code of a Wonder Chemical." Caltech News. California Institute of Technology. Apr. 1980. Volume 14, Number 2. Page 1. http://caltechcampuspubs.library.caltech.edu/2379/1/1980_04_14_02.pdf

constant parade and charade: Ciotti, Paul. "Fighting Disease on the Molecular Front. Leroy Hood Built A Better Gene Machine And The World Beat A Path To His Lab" Los Angeles Times Magazine. 20 Oct. 1985. Page 18. http://articles.latimes.com/1985-10-20/magazine/tm-14242_1_leroy-hood-s-molecular-biology

just a vague new term: Teitelman, Robert. "Gene Dreams: Wall Street, Academia, and the Rise of Biotechnology." Basic Books. A Division of Harper Collins Publishers. 1989

a sensational initial public offering: Genentech IPO. "The Day the World Changed." Life Sciences Foundation. Retrieved 31 July 2014. http://lifesciences-foundation.org/events-Genentech_IPO.html

Biotech was poised for flight: Golden, Frederic. "Shaping Life in the Lab." Time magazine. Time Inc. 9 Mar. 1981. http://bancroft.berkeley.edu/Exhibits/Biotech/Images/6-12lg.jpg

internationally known for his research on the immune system: Caltech Biology Prof Wins Honor. Pasadena Star-News. 27 Mar. 1977. Page A4

you won't find a better communicator: Hendricks, Melissa. "The Big Biology of Leroy Hood." Johns Hopkins Magazine. Feb. 1990

Lee was actually interested in the adventure of ideas: Interview with Roger Perlmutter at the Institute for Systems Biology in Seattle. 24 Jan. 2013

national stature as a scientist: Davidson, Eric letter to Murph Goldberger. Goldberger presidential papers. California Institute of Technology. Caltech Archives. Box 66.11. (Pasadena, CA) Letter dated 13 Apr. 1979.

strongly opposed: Hood, Leroy E. letter to provost Jack Roberts. Goldberger presidential papers. California Institute of Technology Archives. Box 66.11. (Pasadena, CA) Letter dated 10 Feb. 1980.

he was usually traveling around the world: Roberts, John D. Interview by Rachel Prud'homme. Pasadena, California, February 22, 28, March 7, 21, 25, April 12, and May 10, 1985. Oral History Project, California Institute of Technology Archives. Retrieved 4 Dec. 2014 from the World Wide Web: http://resolver. caltech.edu/CaltechOH:OH_Roberts_J

contributions to the Caltech community over the years: Goldberger, Marvin L. Letter to Caltech faculty, students and staff. Goldberger presidential papers. California Institute of Technology Archives. Box 66.11. (Pasadena, CA) Letter dated 24 Mar. 1980.

it is hardly possible to imagine: Zarem, A.M. Letter to Marvin L. Goldberger. 13 May 1982. Provided to the author by Doral Vial.

Caltech was in a difficult time: Simon, Melvin I. Interview by Shirley K. Cohen. Pasadena, California, May 24 and June 5, 2005. Oral History Project, California Institute of Technology Archives. Retrieved 31 July 2014 from the World Wide Web: http://resolver.caltech.edu/CaltechOH:OH_Simon_M

a staggering $400 million: Obituary of Arnold O. Beckman. Caltech News. Engineering & Science No. 2. 2004. http://calteches.library.caltech.edu/4113/1/ Obituaries.pdf

revealed a lack of vision: Interview with Mike Hunkapiller at Pacific Biosciences, Menlo Park, CA. 15 Oct. 2012

impatient as hell: Roberts, John D. Interview by Rachel Prud'homme. Pasadena, California, February 22, 28, March 7, 21, 25, April 12, and May 10, 1985. Oral History Project, California Institute of Technology Archives. Retrieved 4 Dec. 2014 from the World Wide Web: http://resolver.caltech.edu/ CaltechOH:OH_Roberts_J

DuPont nibbled: Cook-Deegan, Robert. "The Gene Wars: Science, Politics, and The Human Genome." W.W. Norton & Co. 1994. Pages 101, 103, 104, 122.

they just don't understand: Interview with Tim Hunkapiller, Mercer Island, WA. 14 Nov. 2012

a plaid shirt, khaki shorts: Interview with Bill Bowes at his San Francisco office. 17 Oct. 2012

you have to make him go shopping: Interview with Eran Hood in Seattle. 29 Nov. 2012

Caltech lawyers weren't just going to give away the store: Fowler, Donald R. Institute Memorandum. "A Chronology of the Interactions with Beckman Representatives RE: the Hood/Hunkapiller Instruments." California Institute of Technology Archives. Goldberger presidential papers. Box 66.11. Memo dated 7 Apr. 1981

Hood was asked to join the scientific advisory board: Hood, Leroy. Foreword to "The

Amgen Story: 25 Years of Visionary Science and Powerful Medicine." Amgen Company History. 2005.

brimming with entrepreneurial excitement and focus: Eletr, Sam. "Twenty Five Years of Advancing Science." Applied Biosystems corporate history. June 2006. http://home.appliedbiosystems.com/about/presskit/pdfs/25_years_advancing_science.pdf

a cancer gene and a natural substance: Schmeck, Harold Jr. "Cancer Gene Tied to Natural Body Substance." The New York Times. 30 June 1983 http://www.nytimes.com/1983/06/30/us/cancer-gene-linked-to-natural-human-substance.html

makes you wonder all sorts of things: Miller, J.A. "Cancer Genes, Growth Factors and the Multi-Step Process." Science News. 18 June 1983. Vol. 123 No. 25. P. 388

Amgen didn't know what to do: Duncan, David E. "The Amgen Story: 25 Years of Visionary Science and Powerful Medicine." Amgen Company History. 2005

founded or co-founded: Leroy Hood biography page. Institute for Systems Biology. Retrieved 21 June 2016 https://www.systemsbiology.org/bio/leroy-hood/

someday, there will probably be a library: Schmeck, Harold Jr. "New Gene Machines Speeding Up the Pace of Biological Revolution." The New York Times. 24 Mar. 1981. Page C1 http://www.nytimes.com/1981/03/24/science/new-gene-machines-speeding-up-the-pace-of-biologial.html

what the cyclotron was to high-energy physics: Ciotti, Paul. "Fighting Disease on the Molecular Front. Leroy Hood Built A Better Gene Machine And The World Beat A Path To His Lab" Los Angeles Times Magazine. 20 Oct. 1985. Page 18. http://articles.latimes.com/1985-10-20/magazine/tm-14242_1_leroy-hood-s-molecular-biology

found the gene: Fu-Kuen Lin, Sidney Suggs, Chi-Hwei Lin, Jeffrey K. Browne, Ralph Smalling, Joan C. Egrie, Kenneth K. Chen, Gary M. Fox, Frank Martin, Zippora Stabinsky, Sayed M. Badrawi, Por-Hsiung Lai, and Eugene Goldwasser. "Cloning and Expression of the Human Erythropoietin Gene." Proceedings of the National Academy of Sciences. Nov. 1985 http://www.pnas.org/content/82/22/7580.full.pdf

first detailed structural characterization of EPO: Lai, Por-Hsiung; Everett, Richard; Wang, Fung-Fang; Arakawa, Tsutomu; and Goldwasser, Eugene. "Structural Characterization of Human Erythropoietin." The Journal of Biological Chemistry. 5 Mar. 1986 http://www.jbc.org/content/261/7/3116.full.pdf

an extraordinary windfall: Krantz, Matt. "Some of the Best Stocks of the 1990s are Some of the Worst of the 2000s." USA Today. 17 Dec. 2009. http://usatoday30.usatoday.com/money/markets/2009-12-17-decadestocks17_ST_N.htm

beating it to death: Interview with Ellen Rothenberg at Caltech. 31 Jan. 2013.

the heart of its being: Baty, Phil. "Caltech: Secrets of the World's Number One University." The Times. 6 Feb. 2014 http://www.timeshighereducation.co.uk/features/caltech-secrets-of-the-worlds-number-one-university/2011008.fullarticle

nobody really reads the literature, broadly: Interview with Irving Weissman at Accelerator Corp. in Seattle. 1 Apr. 2013

we were being affectionate, but snide: Interview with Tim Hunkapiller, Mercer Island, WA. 14 Nov. 2012

let's do that experiment: Interview with Joan Goverman by telephone. 9 May 2014

whatever the boss wanted, he wanted ten minutes ago: Interview with Michael Steinmetz by telephone. 13 Nov. 2012.

you didn't have a snake named Snakey: Interview with Roger Perlmutter at the Institute for Systems Biology in Seattle. 24 Jan. 2013

we all tried to outdo each other: Interview with Ellen Rothenberg at Caltech. 31 Jan. 2013.

Superscientist: Stein, Jeannine. "Superscientist Balances Home Life and Lab Life." Los Angeles Times. (Los Angeles, CA). 14 July 1986. Part V. Page 1. http://articles.latimes.com/1986-07-14/news/vw-19354_1_family-life

She was mortified to see her dad take off his shirt: Interview with Marqui Hood, by telephone. 20 Feb. 2013

the hardest working man in show business: Interview with Roger Perlmutter at the Institute for Systems Biology in Seattle. 24 Jan. 2013

we will learn more: Kevles, Daniel J. and Hood, Leroy E. "The Code of Codes." Harvard University Press. 1992.

CHAPTER 6. "Biology and Medicine in the 21st Century." Page 163. http://www.hup.harvard.edu/catalog.php?isbn=9780674136465

it's transfiguration: Plant, Robert tribute to Elvis Presley. "The 100 Greatest Singers of All-Time." Rolling Stone. 27 Nov. 2008. http://www.rollingstone.com/music/lists/100-greatest-singers-of-all-time-19691231/elvis-presley-20101202

adds to my aura of invincibility: Interview with Mark Davis at Stanford University, Palo Alto, CA. 28 Jan. 2013

Additional sources:

Goldberger, Marvin L. Handwritten letter to Leroy Hood. Goldberger presidential papers. California Institute of Technology Archives. Box 66.11. (Pasadena, CA) Letter dated 31 Oct. 1980.

Faculty of Caltech Biology. Signed petition sent to Leroy Hood, Murph

Goldberger and Rochus Vogt. Goldberger presidential papers. California Institute of Technology Archives. Box 66.10. (Pasadena, CA) Memo dated 18 Apr. 1985.

Goldberger, Marvin L. Letter to Leroy Hood. Goldberger presidential papers. California Institute of Technology Archives. Box 66.11. (Pasadena, CA) 3 May 1985 and 2 Oct 1985.

Goldberger, Marvin L. Handwritten note to Rochus "Robbie" Vogt. Goldberger presidential papers. California Institute of Technology Archives. Box 66.10. (Pasadena, CA) Undated.

CHAPTER 9. COMPETING FOR THE BIG PRIZE

unquestionably the preeminent molecular immunologist: Interview with Roger Perlmutter at the Institute for Systems Biology in Seattle. 24 Jan. 2013

absolute intellectual leader: Interview with Mark Davis at Stanford University, Palo Alto, CA. 28 Jan. 2013

pioneering work in cattle twins: Crow, James F. "A Golden Anniversary: Cattle Twins and Immune Tolerance." Genetics. Genetics Society of America. November 1996. 144: 855-859 http://www.genetics.org/content/144/3/855.full.pdf

a short version: Owen, Ray D. "Immunogenetic Consequences of Vascular Anastomoses Between Bovine Twins." Science. 19 Oct. 1945: 400-401 http://www.sciencemag.org/content/102/2651/400.citation

just as you have 26 letters: Okie, Susan. "Japanese Biologist Wins Nobel Prize; Study of Immune System's Operation Cited." Washington Post. 13 Oct. 1987. Page A3.

Lacking bulletproof data: Weigert ML, Gatmaitan L, Loh E, Schilling L, Hood L. "Rearrangement of Genetic Information may Produce Immunoglobulin Diversity." Nature. 21 Dec. 1978. 276:785-790 http://www.nature.com/nature/journal/v276/n5690/abs/276785a0.html

he wasn't, and still isn't, a technologist: Interview with Mike Hunkapiller at Pacific Biosciences in Menlo Park, CA. 15 Oct. 2012

Tonegawa told his first wife they couldn't have children: Bombardieri, Marcella. "MIT Professor's Drive Draws Awe, Criticism." 7 Aug. 2006. The Boston Globe. http://www.boston.com/news/science/articles/2006/08/07/mit_professors_drive_draws_awe_criticism/?page=full

he was the smartest of all: Dreyer, William. Interview by Shirley K. Cohen. Pasadena, California, February 18–March 2, 1999. Oral History Project, California Institute of Technology

Archives. Retrieved 28 July 2014 from the World Wide Web: http://resolver.caltech.edu/CaltechOH:OH_Dreyer_W

all shared the Lasker Award: Lasker Award Winners Description. The Albert and Mary Lasker Foundation. 1987. Hood, Leroy; Leder, Philip; Tonegawa, Susumu. Retrieved from the World Wide Web, 4 Aug. 2014 http://www.laskerfoundation.org/awards/1987_b_description.htm

much as a railroad controller uncouples and recouples the cars of a freight train: Okie, Susan. Japanese Biologist Wins Nobel Prize; Study of Immune System's Operation Cited. Washington Post. 13 Oct. 1987. Page A3.

we really made a killing: Interview with Mark Davis at Stanford University, Palo Alto, CA. 28 Jan. 2013

what higher tribute can you be paid?: Altman, Lawrence. "MIT Scientist Wins Nobel Prize for Medicine." New York Times. 13 Oct. 1987. Page A1.

I remember … Lee being profoundly excited via Dreyer's thoughts: Interview with Hans Wigzell via email. 13 Nov. 2014

more than the expected number of trips to Sweden: Interview with Mitchell Kronenberg by telephone. 6 June 2014.

many other accomplished scientists and inventors: Staff of National Geographic. "Why Didn't They Win? 10 Huge Discoveries Without a Nobel Prize." 6 Oct. 2014 http://news.nationalgeographic.com/news/2014/10/141006-ten-science-discovery-without-nobel/

he was shocked at how disappointed he was: Interview with Tim Hunkapiller, Mercer Island, WA. 14 Nov. 2012

he felt he deserved it: Interview with Roger Perlmutter at the Institute for Systems Biology in Seattle. 24 Jan. 2013

absolutely pathbreaking: Interview with Ellen Rothenberg at Caltech. 31 Jan. 2013.

one of the central achievements: Weissman, Irving. "Letter to the National Science Foundation Nominating Leroy Hood for the National Medal of Science." 29 Mar. 2011.

the real Holy Grail yet to be found: Weiss, Arthur. "Discovering the TCR Beta-Chain by Subtraction." Journal of Immunology. 1 Sept. 2005. Vol. 175 No. 5 2769-2770 http://www.jimmunol.org/content/175/5/2769.full

we were jolted out of our complacency: Viegas, Jennifer. "Profile of Tak Wah Mak." Proceedings of the National Academy of Science. 11 Nov. 2011. http://www.pnas.org/content/108/48/19124.full

the trio of publications in the same issue: Hedrick, Stephen M.; Nielsen Ellen

A.; Kavaler, Joshua; Cohen, David I.; Davis, Mark. M. "Sequence relationships between putative T-cell receptor polypeptides and immunoglobulins." 8 Mar. 1984. Nature http://www.nature.com/nature/journal/v308/n5955/abs/308153a0.html

the trio of publications in the same issue: Hedrick, Stephen M.; Cohen, David I., Nielsen, Ellen A., Davis, Mark M. "Isolation of cDNA clones encoding T cell-specific membrane-associated proteins." Nature. 8 Mar. 1984 http://www.nature.com/nature/journal/v308/n5955/abs/308149a0.html

the trio of publications in the same issue: Yanagi, Yusuke; Yoshikai, Yasunobu; Leggett, Kathleen; Clark, Stephen P.; I. Aleksander, Ingrid;. Mak, Tak W. "A human T cell-specific cDNA clone encodes a protein having extensive homology to immunoglobulin chains." Nature. 8 Mar. 1984 http://www.nature.com/nature/journal/v308/n5955/abs/308145a0.html

it was over the top, Lee is famously competitive: King, Ralph T. "Gene Machines: An Eclectic Scientist Gives Biotechnology a Fast Assembly Line." Wall Street Journal (New York, NY) 30 May 1995 Page A1

no open desire to crush Davis: Interview with Mitchell Kronenberg by telephone. 6 June 2014.

Additional source:

Mak, Tak W. "The T Cell Antigen Receptor. The Hunting of the Snark." European Journal of Immunology. 31 Oct. 2007. http://onlinelibrary.wiley.com/doi/10.1002/eji.200737443/pdf

CHAPTER 10. THE INVENTION THAT SPARKED A REVOLUTION

arguably one of the most important advances in biology in the 20th century: Pfaelzer, Mariana. "Memorandum of Decision, Findings of Fact, and Conclusions of Law RE Inventorship." U.S. District Court, Central District of California. 19 Feb. 2004.

chart: DeWitt, Natalie D.; Yaffe, Michael P.; Trounson, Alan. "Figure 1: Technology Improvements in DNA Sequencing are Outpacing Moore's Law." Nature Biotechnology. 9 Jan. 2012. http://www.nature.com/nbt/journal/v30/n1/fig_tab/nbt.2086_F1.html http://www.genome.gov/images/content/cost_per_genome2.jpg

a 6-year-old boy from Wisconsin: Johnson, Mark and Gallagher, Kathleen. "One in a Billion: A Boy's Life, a Medical Mystery." Milwaukee Journal-Sentinel. Three-part series. 18 Dec. 2010, 21 Dec. 2010, 25 Dec. 2010. http://www.jsonline.com/news/health/111641209.html http://www.jsonline.com/news/health/112248249.html http://www.jsonline.com/news/health/112249759.html

Glauber, Bill. "Journal Sentinel Wins Pulitzer Prize for 'One in a Billion' DNA Series." 18 Apr. 2011 http://www.jsonline.com/news/milwaukee/120091754.html

other desperately ill children: Mnookin, Seth. "One of a Kind." The New Yorker. 21 July 2014. http://www.newyorker.com/magazine/2014/07/21/one-of-a-kind-2

hopeful narrative spread: Cracking Your Genetic Code. Nova special. PBS 29 Mar. 2012. http://www.pbs.org/wgbh/nova/body/cracking-your-genetic-code.html

his vision and *the 1980s appear to be the decade:* Hood, Leroy and Hunkapiller, Michael. "Biotechnology and Medicine of the Future." Engineering and Science. California Institute of Technology. March 1983. http://calteches.library.caltech.edu/560/2/Hunkapiller.pdf

the microchemical facility: Hunkapiller, Michael; Hood, Leroy; et al "A Microchemical Facility for the Analysis and Synthesis of Genes and Proteins." Nature. 12 July 1984. http://www.nature.com/nature/journal/v310/n5973/abs/310105a0.html

he had developed phosphorus chemistry methods: Matteucci, Mark, and Caruthers, Marvin H. "Synthesis of deoxyoligonucleotides on a polymer support." Journal of the American Chemical Society. 1981 http://pubs.acs.org/doi/abs/10.1021/ja00401a041

he had developed phosphorus chemistry methods: Beaucage, Serge, and Caruthers, Marvin H. "Deoxynucleoside phosphoramidites–a new class of key intermediates for deoxypolynucleotide synthesis." Tetrahedron Letters. 1981 http://www.sciencedirect.com/science/article/pii/S0040403901904617

why bother with automation: Hood, Lee. "A Personal Journey of Discovery: Developing Technology and Changing Biology." Annual Review of Analytical Chemistry. Annual Reviews. (Palo Alto, CA). 20 Feb. 2008 http://www.ncbi.nlm.nih.gov/pubmed/20636073

didn't publish anything on their work until years later: Horvath, Suzanna J., Firca JR, Hunkapiller, Tim J., Hunkapiller Michael W., Hood, Leroy E. (1987) An Automated DNA Synthesizer Employing Deoxynucleoside 3' Phosphoramidites. Methods in Enzymology. 1987 http://www.ncbi.nlm.nih.gov/pubmed/3481013

wrote that Sam Eletr: Carruthers, Marvin H. "The Chemical Synthesis of DNA/RNA: Our Gift to Science." The Journal of Biological Chemistry. 11 Jan. 2013 http://www.jbc.org/content/288/2/1420.full.pdf

Hood's group tried—largely unsuccessfully—to automate an early synthesis chemistry from Marv: Hunkapiller, Michael. Interview by e-mail. 29 Sept. 2014

this prejudice against technology was ridiculous: Interview with Maynard Olson at the University of Washington. 29 Nov. 2012.

other things were duck soup by comparison: Hood, Leroy E. Interviewed by Arnold

Thackray, David C. Brock, and Arthur Daemmrich. New Orleans, Louisiana. 19 Mar. 2002. Beckman Heritage Project, Chemical Heritage Foundation.

Smith was only dimly aware of Hood's reputation: Interview with Lloyd Smith at the University of Wisconsin. 6 Dec. 2012

drove a beat up 1964 Chevy Nova and *I found myself in the trenches:* Smith, Lloyd M. "The Development of Automated DNA Sequencing." Undated, unpublished manuscript. Provided to the author. 6 Dec. 2012.

the necessary chemistry has been developed: Smith, Lloyd M.; Hunkapiller, Michael W.; Hunkapiller, Tim J.; "Invention Disclosure Form for Automated DNA Sequencing Machine." California Institute of Technology (Pasadena, CA).

a meeting on Oct. 1, 1982: Conception of invention dated 1 Oct. 1982. First written description of invention 29 Oct. 1982. Form filing date, 28 Nov. 1983. Witnessed by Hood, Leroy.

absolutely remarkable progress: Hood, Lee. "Letter to Richard N. Mackay, president of the Donald E. Baxter Foundation." 19 Nov. 1984.

a small but influential technical journal: Smith, Lloyd M; Fung, Steven; Hunkapiller, Michael W.; Hunkapiller, Tim J.; Hood, Leroy E. "The Synthesis of oligonucleotides containing an analiphatic amino group at the 5' terminus; synthesis of fluorescent DNA primers for use in DNA sequence analysis." Nucleic Acids Research. 11 Apr. 1985 http://smith.chem.wisc.edu/content/synthesis-oligonucleotides-containing-aliphatic-amino-group-5-terminus-synthesis-fluorescent http://www.ncbi.nlm.nih.gov/pubmed/4000959

paper describing the first automated DNA sequencer: Smith, Lloyd M.; Sanders, Jane Z.; Kaiser, Robert J; Hughes, Peter; Dodd, Chris; Connell, Charles L.; Heiner, Cheryl; Kent, Stephen B.H.; Hood, Leroy E. "Fluorescence Detection in Automated DNA Sequence Analysis." Nature. 12 June 1986. http://www.nature.com/nature/journal/v321/n6071/abs/321674a0.html

this machine will automate many of the tasks: Caltech Scientists Invent Automatic DNA Sequencer. Caltech press release. 17 July 1986.

In a development expected to further accelerate: Maugh, Thomas H. "Caltech Scientists Develop Super-Fast DNA Analyzer." Los Angeles Times. 12 June 1986. http://articles.latimes.com/1986-06-12/news/mn-10208_1_dna-sequence

On inspection, I could not find any automation: Church, George M. Draft book foreword sent to the author. 12 June 2014.

should I come back for the press conference: Interview with Lloyd Smith at the University of Wisconsin. 6 Dec. 2012

we were so mad, ABI was mad: Interview with Tim Hunkapiller, Mercer Island, WA. 14 Nov. 2012 *went out of his way to say:* Hunkapiller, Michael. Fax sent from Applied Biosystems to Robert Cook-Deegan. 18 Mar. 1991.

this was my future in a crate: Venter, J. Craig. "A Life Decoded. My Genome: My Life." Pg. 100-101. Viking. 2007. http://www.amazon.com/ Life-Decoded-My-Genome/dp/0143114182

the Hood group wasn't even regularly using its own heralded machine: Hunkapiller, Michael. E-mailed response to questions. 21 Nov. 2014

I assembled a team: Hood, Lee. "A Personal Journey of Discovery: Developing Technology and Changing Biology." Annual Review of Analytical Chemistry. Annual Reviews. (Palo Alto, CA). 20 Feb. 2008 http://www.ncbi.nlm.nih.gov/ pubmed/20636073

in that dystopian book: Crichton, Michael. "Jurassic Park: A Novel." Random House Publishing Group. 1991. Page 39. http://www.jplegacy.org/ jpencyclopedia/?p=2897

that's simply not correct: Maugh, Thomas H. "A look inside Hollywood and the movies : Dinologists Aren't Saur, but 'Jurassic Park' Gets DNA Docs' Amber Up." Los Angeles Times. 20 June 1993. http://articles.latimes.com/1993-06-20/ entertainment/ca-5004_1_jurassic-park

no single person: Jones, Mark; Dick, Brian; Nelson, Heather. "Honoring 25 Years of Biotech Leadership. The Biotech Hall of Fame Awards." Life Sciences Foundation. 2012. Page 23. http://lifesciencesfoundation.org/news-lsf_pub-lishes_book_honoring_biotech_leaders.html

the expose-style headline: Gosselin, Peter G. and Jacobs, Paul. "DNA Device's Heredity Scrutinized by U.S.; Scientists Insist They Invented Technique Without Federal Funding, But Paper Trail Suggests They Relied on Millions in Grants. Inquiry Will Determine If There is Need for Payback." Los Angeles Times. 14 May 2000. http://articles.latimes.com/2000/may/14/news/mn-30009

later at USA Today: Eisler, Peter. "Was Government Cheated on DNA Decoder?" USA Today. 18 Apr. 2002. http://usatoday30.usatoday.com/news/science/ biology/2002-04-18-cover-genome.htm

biologists who once had to kill 1,000 mice: Dietrich, Bill. "Future Perfect -- Thanks To Bill Gates' $12-Million Endowment, Scientist Leroy Hood Continues His Search For A New Genetic Destiny." The Seattle Times. Pacific Magazine. 9 Feb. 1992. Page 6. http://community.seattletimes.nwsource.com/archive/?date=1992 0209&slug=1474735

work that created several new scientific disciplines: Remarks by President Obama at the White House ceremony, presenting the National Medal of Science. 1 Feb. 2013 http://www.whitehouse.gov/the-press-office/2013/02/01/ remarks-president-ceremony-2011-national-medals-science-and-national-med

I see every day the impact of Lee's contributions: Gates, Bill. "Letter of Nomination for the National Medal of Science." 14 Mar. 2012

More than $100 million in royalties were collected: Interview with Larry Gilbert at Caltech campus in Pasadena, CA. 1 Feb. 2013

CHAPTER 11. THE CLASH AT CALTECH

Sinsheimer had just gotten a $35 million gift: Sinsheimer, Robert "The Santa Cruz Workshop – May 1985." Genomics. Volume 5, Issue 4, Nov. 1989, Pages 954–956 http://www.sciencedirect.com/science/article/pii/0888754389901420

bold, exciting—but simply not feasible: Roberts, Leslie. "Controversial from the Start" Science. 16 Feb. 2001. http://science.sciencemag.org/content/291/5507/news-summaries

do we get the money?: Cook-Deegan, Robert. "The Gene Wars: Science, Politics, and The Human Genome." W.W. Norton & Co. 1994. Pages 101, 103, 104, 122.

a series of essays: Kevles, Daniel J. and Hood, Leroy E. "The Code of Codes." Harvard University Press. 1992. Preface pages vii-viii.

alarm, derision, and outright fury: Angier, Natalie. "Great 15-Year Project to Decipher Genes Stirs Opposition." The New York Times. 5 June 1990. http://www.nytimes.com/1990/06/05/science/great-15-year-project-to-decipher-genes-stirs-opposition.html?src=pm&pagewanted=1

he's riding with the tide: Benzer, Seymour. Interview by Heidi Aspaturian. Pasadena, California, September 11-February 1991. Oral History Project, California Institute of Technology Archives. Retrieved 3 Nov. 2014 from the World Wide Web: http://resolver.caltech.edu/CaltechOH:OH_Benzer_S

evangelical enthusiasm of the modern Knights Templar: Davies, Kevin. "Cracking the Genome: Inside the Race to Unlock Human DNA." Johns Hopkins University Press. 2002. Page 12 http://www.amazon.com/Cracking-Genome-Inside-Unlock-Human/dp/0801871409

Everything!: Simon, Melvin I. Interview by Shirley K. Cohen. Pasadena, California, May 24 and June 5, 2005. Oral History Project, California Institute of Technology Archives. Retrieved 1 Aug. 2014 from the World Wide Web: http://resolver.caltech.edu/CaltechOH:OH_Simon_M

scored a hefty: Interview with Eric Davidson at his Caltech office. Pasadena, CA. 31 Jan. 2013.

when the facilities group: Interview with Ellen Rothenberg at her Caltech office. 31 Jan. 2013.

an astonishing 74 research projects: A Report for the Year 1987-1988 on the

Research and Other Activities of the Division of Biology at the California Institute of Technology.

has a tendency to pick up strays: Interview with Roger Perlmutter at the Institute for Systems Biology in Seattle. 24 Jan. 2013

we were scientifically mediocre: Interview with Joan Goverman by telephone. 9 May 2014

I was amazed: Interview with Thomas Everhart by telephone. 22 Jan. 2013

you ask a lot of questions: Thomas E. Everhart Named Caltech President. Caltech News. California Institute of Technology. Aug. 1987. Page 1. http://caltechcampuspubs.library.caltech.edu/2421/1/1987_08_21_04.pdf

won a big grant: Caltech a Winner in NSF Competition. Caltech News. California Institute of Technology (Pasadena, CA). Volume 23, Issue 1, Page 1. http://caltechcampuspubs.library.caltech.edu/2437/1/1989_02_23_01.pdf

a secret cabal: Interview with Eric Davidson at his Caltech office. Pasadena, CA. 31 Jan. 2013.

everything Lee tried to do was blocked: Interview with Ellen Rothenberg at her Caltech office. 31 Jan. 2013.

he inquired at UCLA: Hood, Leroy E. Letter to Phelps, Michael E. 21 May 1990.

dangled the possibility: Hood, Leroy E. Letter to Shine, Kenneth I. 14 June 1990.

interest fizzled: Interview with John Yates by telephone. 12 Nov. 2012

move heaven and earth: Interview with Daniel Kevles by telephone 3 Nov. 2014

the most agonizing decision of my life: Hood, Leroy E. Letter to Gerald Rubin at UC Berkeley. 21 Dec. 1990

everyone in the lab breathed an enormous sigh of relief: Interview with Lee Rowen at the Institute for Systems Biology in Seattle. 19 Dec. 2012.

an extremely productive stay in my laboratory: Hood, Leroy E. Letter of recommendation for Vipin Kumar, sent to Charles Earhart at the University of Texas. 16 Apr. 1990

Kumar's paper: Kumar, Vipin; Urban, James L.; Hood, Leroy E. "In Individual T Cells One Productive Alpha Rearrangement Does Not Appear to Block Rearrangement at the Second Allele." Journal of Experimental Medicine. Rockefeller University Press. Dec. 1989.

a little more than a month later: Hood personal files. "Kumar List of Chron Events 1990-1991"

a painful public spectacle: Kevles, Daniel J. "The Baltimore Case: A Trial of Politics, Science, and Character." W.W. Norton & Co. 1998.

had to act fast—and with a certain sensitivity: Cassata, Donna. "NIH Researcher Commits Suicide While Under Pressure From Investigation." The Associated Press. 14 Feb. 1987 http://www.apnewsarchive.com/1987/NIH-Researcher-Commits-Suicide-While-Under-Pressure-From-Investigation/id-1eff7637a702be2bf0097bf849d50d7f

working in the dark: Roberts, Leslie. "Misconduct: Caltech's Trial by Fire." Science. 20 Sept. 1991. http://science.sciencemag.org/content/253/5026/1344

he did listen to reason: Interview with Joan Goverman by phone. 9 May 2014

Valerie picked up Kumar: Hood personal files. "Kumar List of Chron Events 1990-1991"

imperative that the integrity of science and our institution be maintained: Hood, Leroy E. Letter to the Caltech faculty and staff. 13 June 1990.

a vigorous effort to examine the data: Hood personal files. Jennings statement to faculty re Kumar issue, 1990.

the NIH investigation of Kumar: Findings of Scientific Misconduct. National Institutes of Health. 5 July 1996. http://grants.nih.gov/grants/guide/notice-files/not96-177.html

Urban had written a paper: Goodstein, David. "On Fact and Fraud: Cautionary Tales From the Front Lines of Science." Princeton University Press. 2010.

Urban accepted a three-year ban: Findings of Scientific Misconduct. National Institutes of Health. 28 July 1995. http://grants.nih.gov/grants/guide/notice-files/not95-181.html

Despite Hood's objections, Science ran the article: Roberts, Leslie. "Caltech Deals with Fraud Allegations." Science. 1 Mar. 1991 http://www.ncbi.nlm.nih.gov/pubmed/1998111

the innocent suffer: Interview with Ellen Rothenberg at her Caltech office. 31 Jan. 2013.

he would have been a terrible politician: Interview with Eran Hood in Seattle. 29 Nov. 2012

CHAPTER 12. A FRESH START IN SEATTLE

the city still hadn't entirely recovered: City of Seattle Department of Planning and Development http://www.seattle.gov/dPd/cityplanning/populationdemographics/default.htm

you can't ruin a splendid university in five or six years: Balta, Victor. "William P. Gerberding, 27th president of the University of Washington, dies at 85."

UW Today. 29 Dec. 2014. http://www.washington.edu/news/2014/12/29/william-p-gerberding-27th-president-of-the-university-of-washington-dies-at-85/

Perlmutter moved from Caltech to Seattle: Interview with Roger Perlmutter at the Institute for Systems Biology in Seattle. 24 Jan. 2013

let's create a whole new department: Interview with Lee Huntsman in Seattle. 26 Nov. 2012

Dr. Hood championed a research program: Danz lecture recordings. Apr. 1991. Courtesy of University of Washington archives.

He's a very brilliant guy: Interview with Bill Gates, responses in writing, 25 Oct. 2014

When you first meet Lee,: Lazowska, Ed. Video interview played at Hood's 70th birthday event at the W Hotel in downtown Seattle, 10 Oct. 2008

an awesome and at times frightening blend of brilliance, drive, competitiveness: Isaacson, Walter. "In Search of the Real Bill Gates." Time magazine. 24 June 2001. http://content.time.com/time/magazine/article/0,9171,137132,00.html

we are keen to have Lee: Perlmutter, Roger. "Memo to Phil Fialkow." 24 May 1991. University of Washington Special Collections. UW Department of Genome Sciences. Accession No. 06--031

Bill struck me as an individual: Hood, Leroy "Handwritten memo to Phil Fialkow" Undated, 1991. University of Washington Special Collections. UW Department of Genome Sciences. Accession No. 06—031

divvied up several ways: Hadley, Jane. "Gates Gives $12 million to UW Venture." Seattle Post-Intelligencer. 8 Oct. 1991. Page A1.

you don't find this parochial view in physics: Interview with Maynard Olson at the University of Washington Department. 29 Nov. 2012.

sad duty: Hood, Leroy E. Letter to Arnold O. Beckman. 7 Oct. 1991

a mop of hair: Neves, Randy. "Gates Gives Record $12 million." The Daily. University of Washington. 8 Oct. 1991. Page 1.

bankrolled a major coup: Bishop, Jerry E. "William Gates's Gift Spurs Shift by Star Biologist." Wall Street Journal. 8 Oct. 1991. Page B1

we hope to change the world: Macdonald, Sally. "With Gates' Bait, UW Lands a Big Fish in the Gene Pool." The Seattle Times. 8 Oct. 1991. Page B4 http://community.seattletimes.nwsource.com/archive/?date=19911008&slug=1309734

the glow over this institution: Gerberding, William P. "Thank you letter to Bill Gates." 14 Oct. 1991. University of Washington Special Collections. UW Department of Genome Sciences. Accession No. 06—031

complete pandemonium: Interview with John Yates by telephone. 12 Nov. 2012

not particularly drawn to Seattle: Interview with Lee Rowen at the Institute for Systems Biology in Seattle. 19 Dec. 2012.

University of Washington is not among the top five: Dickinson, Susan L-J., "21st Century Facility is Bound for Seattle." The Scientist. Page 1. 30 Mar. 1992.

pitched his vision: Hood, Leroy; Ramsey, Paul; Stamatoyannopoulos, George. "Offer letter to Maynard Olson." 18 Dec. 1991. University of Washington Special Collections. Accession #14—001

we got everyone we wanted: Roberts, Leslie. "Leroy Hood: Thinking Big in Seattle." Science. 8 Apr. 1994. http://www.sciencemag.org/content/264/5156/206.extract

the cover of the Seattle Times Sunday magazine: Dietrich, Bill. "Future Perfect -- Thanks To Bill Gates' $12-Million Endowment, Scientist Leroy Hood Continues His Search For A New Genetic Destiny." The Seattle Times. Pacific Magazine. 9 Feb. 1992. Page 6. http://community.seattletimes.nwsource.com/archive/?date=19920209&slug=1474735

sheer luck: Sandoz Buying Systemix' Stem Cell and Megakaryocyte Technology for $392 Mil.; Deal Values Firm at $650 Mil-$700 Mil. – Leading Edge Technology Benchmark? The Pink Sheet. 23 Dec. 1991 http://www.pharmamedtechbi.com/publications/the-pink-sheet/53/051/sandoz-buying-systemix-stem-cell-and-megakaryocyte-technology-for--392-mil-deal-values-firm-at http://www.nytimes.com/1991/12/18/business/business-people-systemix-sandoz-accord-linked-to-scientist-s-role.html

doing the dishes, on his own: Interview with Carl Weissman at Accelerator Corp. in Seattle. 17 Dec. 2012

some biologists are so into the minutiae: Interview with Nathan Myhrvold at Intellectual Ventures. 27 Nov. 2012

Melinda French: Monday Memo. "Out of Africa." The Seattle Times 27 Sept. 1993 http://community.seattletimes.nwsource.com/archive/?date=19930927&slug=1723151

such a good teacher: Interview with Ann Winblad at Hummer Winblad offices, San Francisco. 8 Jan. 2013

CHAPTER 13. NEW FRONTIERS, BOLD PROMISES

nobody could accuse Leroy Hood of thinking small: Roberts, Leslie. "Leroy Hood: Thinking Big in Seattle." Science. 8 Apr. 1994. http://www.sciencemag.org/content/264/5156/206.extract

Hood predicted that by 2016: Garrett, Laurie. "The Dots Are Almost Connected.... Then What?; Mapping the Human Genetic Code; They've Almost Cracked the Human Genetic Code. That's the Good News. What Happens After That is the Harder Part." Los Angeles Times Magazine. 3 Mar. 1996. http://articles.latimes.com/1996-03-03/magazine/tm-42636_1_human-gene

first complete genome of a free-living organism: Sequencing the genome of Haemophilus influenzae Rd. Genome News Network. http://www.genomenews-network.org/resources/timeline/1995_Haemophilus.php

eliminate the need for prior mapping: Roach, Jared C., Boysen, Cecilie, Wang, Kai, Hood, Leroy E. "Pairwise end sequencing: a unified approach to genomic mapping and sequencing." Genomics. 20 Mar. 1995 http://www.sciencedirect.com/science/article/pii/088875439580219C

smashing: Interview with Nathan Myhrvold at Intellectual Ventures in Bellevue, WA. 27 Nov. 2012

a blockbuster: Fleischmann RD et al. Whole-genome random sequencing and assembly of Haemophilus influenzae Rd. Science. 28 July 1995 http://www.ncbi.nlm.nih.gov/pubmed/7542800

we used to joke: Interview with Lee Huntsman in Seattle. 26 Nov. 2012

creature of the university: Interview with Maynard Olson at the University of Washington. 29 Nov. 2012

you didn't need to be in the room five minutes: Interview with Joan Goverman by phone. 9 May 2014

see what we could do in a few years, not a few decades: Haney, Daniel Q. "Michael Milken Fights Cancer." The Associated Press. 25 Nov. 1995

in the zone: Interview with Stephen Friend at the Fred Hutchinson Cancer Research Center. 17 Dec. 2012

I'd be shocked: King, Ralph T. Jr. "Gene Machines: An Eclectic Scientist Gives Biotechnology a Fast Assembly Line." Wall Street Journal. 30 May 1995 Page A1

so unsafe as to be absurd: Church, George M. Draft book foreword sent to the author. 12 June 2014.

willing to listen and learn from us: Interview with Jane Butler Kahle, via email. 9 Sept. 2014

many evenings at school board meetings: Interview with Dana Riley Black at the Institute for Systems Biology. 18 Dec. 2012

the Seattle scientist who has been most energetic: Dietrich, Bill; Long, Katherine; Lilly, Dick; Beason, Tyrone; Houtz, Jolayne. "How Do You Spell Success? S-C-I-E-N-C-E." The Seattle Times 7 Apr. 1996 http://community.seattletimes.

nwsource.com/archive/?date=19960407&slug=2323073

it was kind of a dicey field: Interview with Daniel Kevles by telephone. 3 Nov. 2014

completely spellbound: Interview with Trey Ideker by telephone. 18 Dec. 2012

wasn't going into a department with Lee: Interview with Alan Aderem at the Center for Infectious Disease Research in Seattle. 3 Jan. 2013

Lee is like the sun: Interview with Roger Perlmutter at the Institute for Systems Biology 24 Jan. 2013

tragedy struck: Vacationing UW Medical School Dean, Wife and Guides Perish in Nepal. December 1996. UW Columns. https://www.washington.edu/alumni/columns/dec96/fialkow.html

CHAPTER 14. A MESSY DIVORCE

touting this external validation: Molecular Dynamics press release. Jan. 12, 1998. http://www.siliconinvestor.com/readmsg.aspx?msgid=3146817

accounting was a nightmare: University of Washington Special Collections. Internal records, Department of Molecular Biotechnology/Genome Sciences

Patterson, Harlan. "Memo to Bob Gust." 26 Sept. 1994. University of Washington Special Collections. Accession #14—001

message wasn't moored in the messy realities: Interview with Maynard Olson at the University of Washington. 29 Nov. 2012

the NIH totally shit-canned that grant: Interview with Lee Rowen at the Institute for Systems Biology in Seattle. 19 Dec. 2012.

an influential editorial: Olson, Maynard. "A Time to Sequence." Science. 20 Oct. 1995 http://www.sciencemag.org/site/feature/data/genomes/270-5235-394.pdf

the Olson camp and the Hood camp: Interview with Todd Smith in Seattle. 4 Jan. 2013

Olson, Maynard. "Memo to Paul Ramsey." 24 July 1996. University of Washington Special Collections. Accession #14—001

Olson, Maynard. "Memo to Paul Ramsey, Bob Gust, Leroy Hood, Phil Green." 13 Sept. 1996. University of Washington Special Collections. Accession #14—001

the battle lines were drawn: Varmus, Harold. "The Art and Politics of Science." W.W. Norton & Co. 2009. Ch. 10. Being NIH Director. http://www.ncbi.nlm.nih.gov/books/NBK190624/

Darth Venter: I'm the Human Genome. McKie, Robin. The Guardian. 28 Apr. 2002. https://www.theguardian.com/world/2002/apr/28/medicalscience.research

Hood is kind of like Craig Venter, but with real accomplishments: Interview with Maynard Olson at the University of Washington Foege Hall, Department of Genome Sciences. 29 Nov. 2012

advocating large-scale shotgun sequencing: Venter, J. Craig; Smith, Hamilton O.; Hood, Leroy E. "A new strategy for genome sequencing." Nature 30 May 1996. http://www.ncbi.nlm.nih.gov/pubmed/8632789

we can do this on any large genome: Interview with Greg Mahairas in Seattle. 16 Nov. 2012

rice genome was important: Eckhardt, Nancy A. "Sequencing the Rice Genome." 12 Nov. 2000 Plant Cell. http://www.ncbi.nlm.nih.gov/pmc/articles/PMC526008/

unique opportunity for Monsanto: Hood, Leroy E. Letter to Monsanto CEO Robert Shapiro. 7 July 1998

ought to be bloody appreciative: Huntsman, Lee. "E-mail message to Paul Ramsey." 31 Oct. 1997. University of Washington Special Collections. Accession #14—001

wondering if there is a shot at setting you up back here: Baltimore, David. Letter to Leroy Hood. 24 Feb. 1998

The feeling, particularly in biology, was so strong against him: Interview with David Baltimore at his Caltech office. 30 Jan. 2013

particularly horrifying: Hood, Leroy. Letter to Bob Gust on "The Institute for the Analysis of Molecular Systems." 25 June 1998

we have been timely and generous: Ramsey, Paul. "Memo to Leroy Hood." 15 Oct. 1998. University of Washington Special Collections. Accession #14—001

someone who could be a peacemaker: Interview with Roger Perlmutter at the Institute for Systems Biology in Seattle. 24 Jan. 2013

Pollack, Andrew "A Biotech Superstar Looks at the Bigger Picture." The New York Times. 17 Apr. 2001. Page F3 http://www.nytimes.com/2001/04/17/science/scientist-at-work-leroy-hood-a-biotech-superstar-looks-at-the-bigger-picture.html

getting started with an anonymous $5 million donation: Lane, Polly; Sanchez, Roberto. "Geneticist Leroy Hood Leaves UW – He'll Lead Private Venture to Link Computer Science With Biology." The Seattle Times. 13 Dec. 1999 http://community.seattletimes.nwsource.com/archive/?date=19991213&slug=3001252

Hood, Leroy; Aderem, Alan; Aebersold, Ruedi. "History of the Institute for Systems Biology." Unpublished internal copy provided by Leroy Hood.

I pleaded with him: Kuzma, Cindy. "The Pied Piper of Science." Science & Spirit. Quincy, Massachusetts. January February 2004. Page 26. http://cindykuzma.files.wordpress.com/2011/02/the-pied-piper-of-science.pdf

oh my God: Interview with Eran Hood in Seattle 29 Nov. 2012

it was personal: Interview with Alan Aderem. 3 Jan. 2013

he's a wonderful guy, but a very demanding guy: Interview with Bill Gates, responses in writing, 25 Oct. 2014

frog jumping into a frying pan: Comments via email from Todd Smith, 9 Jan. 2015.

CHAPTER 15. START-UP DAYS

Hood, Leroy. Internal strategy document for Institute for Systems Biology. January 2000. Provided by Hood

Shares of Amgen rose 5,788 percent in the 1990s: Krantz, Matt. "Some of the Best Stocks of the 1990s Are Worst in 2000s." USA Today. 17 Dec. 2009 http://usa-today30.usatoday.com/money/markets/2009-12-17-decadestocks17_ST_N.htm

investigation by the inspector general of the US Department of Health and Human Services: Gosselin, Peter G.; Jacobs, Paul. "U.S. Officials Probe Cost of Genetic Decoder." Los Angeles Times. 16 Feb. 2000. http://articles.latimes.com/2000/feb/16/news/mn-64996

it's a witch-hunt: Cohen, Jon; Marshall, Eliot; Roberts, Leslie; Pennington, Elizabeth. "HHS Probes Genesis of Gene Sequencer." Science. American Association for the Advancement of Science 25 Feb. 2000. http://www.sciencemag.org/content/287/5457/1374.2.summary?sid =13a95642-1a62-4d20-9c07-99b496ff61e2

investigative follow-up: Gosselin, Peter G.; Jacobs, Paul. "DNA Device Heredity Scrutinized by US" Los Angeles Times. 14 May 2000. http://articles.latimes.com/2000/may/14/news/mn-30009

he is a rock: Interview with Alan Aderem at Center for Infectious Disease Research. 3 Jan. 2013.

chief cook and bottle washer: Interview with Louis Coffman at Center for Infectious Disease Research in Seattle. 24 Jan. 2013.

there was no going back: Interview with Ruedi Aebersold from his office in Switzerland, via Skype. 20 Dec. 2012

knew I wanted him to be my adviser: Interview with Trey Ideker via telephone. 18 Dec. 2012

McCormick offered some diplomatic comments: The Chronicle of Higher Education. 14 Jan. 2000. Page A14

a humane and conscientious boss: Interview with Tawny Biddulph Burns by telephone. 12 June 2014

if you could do anything you wanted, what would you do?: Interview with John Aitchison at Center for Infectious Disease Research in Seattle. 8 July 2014

a manifesto of systems biology: Ideker, Trey; Galitski, Tim; Hood, Leroy. "A New Approach to Decoding Life: Systems Biology." Annual Review of Genomics and Human Genetics. http://idekerlab.ucsd.edu/Documents/idekerGH2001.pdf

especially important for a new institute striving for peer validation: Ideker, Trey; Thorsson, Vesteinn; Ranish, Jeffrey A.; Christmas, Rowan; Buhler, Jeremy; Eng, Jimmy K.; Bumgarner, Roger; Goodlet, David R.; Aebersold, Ruedi; Hood, Leroy. "Integrated Genomic and Proteomic Analyses of a Systematically Perturbed Metabolic Network." Science. 4 May 2001 http://www.sciencemag.org/content/292/5518/929.short

this will cement the leadership role the university is in: Timmerman, Luke. "Gates Gives $70 Million for Genome Work at UW." The Seattle Times. 24 Apr. 2003. Page A1 http://community.seattletimes.nwsource.com/archive/?date=20030424&slug=gatesgift24

Lee has built an impressive team: Press release. Institute for Systems Biology. 19 Oct. 2005. http://www.systemsbiology.org/Press_Release_101905

We love Lee's work, but it's not in the foundation's sweet spot: Interview with Bill Gates, responses in writing, 25 Oct. 2014

horribly exposed: Interview with Louis Coffman at Center for Infectious Disease Research in Seattle. 24 Jan. 2013.

considerable skepticism: Pollack, Andrew "A Biotech Superstar Looks at the Bigger Picture." The New York Times. 17 Apr. 2001. Page F3 http://www.nytimes.com/2001/04/17/science/scientist-at-work-leroy-hood-a-biotech-superstar-looks-at-the-bigger-picture.html

impact of its scientific publications: Timmerman, Luke. "ISB Wins Top Scientific Impact Rank." Xconomy. 7 Jan. 2010 http://www.xconomy.com/seattle/2010/01/07/isb-wins-top-scientific-impact-rank/

heroes of manufacturing: Bylinsky, Gene. "Lee Hood: The Man Who Automated Biology." Fortune. 17 Mar. 2003. http://archive.fortune.com/magazines/fortune/fortune_archive/2003/03/17/339245/index.htm

golden publicity: Medicine Without Frontiers. The Economist. 15 Sept. 2005. http://www.economist.com/node/4368242?zid=314&ah=607477d0cfcfcoadb6dd off57bb8e5c9

some even say it's impossible: Schonfeld, Erick. "Beyond the Genome." Business 2.0. July 2003. http://money.cnn.com/magazines/business2/business2_archive/2003/07/01/345245/index.htm

role models for youth of America: Berger, Eric. "Who are America's Science Role

Models? It's Ugly Folks." Houston Chronicle. 21 Mar. 2008. http://blog.chron.com/sciguy/2008/03/who-are-americas-science-role-models-its-ugly-folks/

scientific illiteracy: Mooney, Chris and Kishenbaum, Sheril. "Unscientific America: How Science Illiteracy Threatens Our Future." 2009. Basic Books. http://www.amazon.com/gp/product/0465013058/

CHAPTER 16. A VISIONARY, NOT A MANAGER

we didn't want to go on living as if we were vegetables: Interview with Eric Adelberger by telephone. 8 Apr. 2014

handles so many things himself: Interview with Marqui Hood by phone. 20 Feb. 2013

the sun has shone brightly on you: Interview with Louis Coffman at Center for Infectious Disease Research in Seattle. 24 Jan. 2013.

how the fuck is this an opportunity?: Interview with Alan Aderem at Center for Infectious Disease Research. 3 Jan. 2013.

planned to run the institute four to six more years: Timmerman, Luke. "Creating a Hothouse for Bioscience Research." The Seattle Times. (Seattle, WA). 25 Apr. 2002. http://community.seattletimes.nwsource.com/archive/?date=20020425&slug=hood25

doing the most ambitious things: Timmerman, Luke. "Leroy Hood, Turning 70, Still Aims to Accomplish 'The Most Ambitious Things of My Career." Xconomy. 10 Oct. 2008. http://www.xconomy.com/seattle/2008/10/10/leroy-hood-turning-70-still-aims-to-accomplish-the-most-ambitious-things-of-my-care

your energy is amazing: Video of Bill Gates displayed at Hood's 70th birthday party, W Hotel, Seattle, 10 Oct. 2008

the guy's a maniac: Video of Alan Aderem displayed at Hood's 70th birthday party, W Hotel, Seattle, 10 Oct. 2008

Hood usually thought that was unnecessary: Interview with Larry Herron in Seattle. 15 Feb. 2013

Luxembourg bit: Leading U.S. Bioscience Pioneers Enter International Collaboration With Government of Luxembourg to Accelerate Biomedical Research. Press release. Institute for Systems Biology. 6 June 2008. http://www.systemsbiology.org/Press_Release_60608

he sees the future very clearly, sometimes he isn't so good in the present: Interview with Roger Perlmutter at the Institute for Systems Biology in Seattle. 24 Jan. 2013 *do-gooder attitudes:* Interview with Eric Davidson at his Caltech office. 31 Jan. 2013.

Aderem's exit was a body blow: Timmerman, Luke. "Alan Aderem, With Team in Tow, Bolts From ISB to Take Leading Role at Seattle BioMed." Xconomy. 14 Mar. 2011 http://www.xconomy.com/seattle/2011/03/14/alan-aderem-with-team-in-tow-bolts-from-isb-to-take-leading-role-at-seattle-biomed/

Luxembourg officials had gotten what they wanted: Interview with Rudi Balling via email. 8 Sept. 2014

the P4 thing, it's such bullshit: Interview with Mark Davis at Stanford University, Palo Alto, CA. 28 Jan. 2013

usefulness of sequencing a whole family: Roach, JC et al, Hood, LE. "Analysis of genetic inheritance in a family quartet by whole-genome sequencing." *Science.* 30 Apr. 2010 http://www.ncbi.nlm.nih.gov/pubmed/20220176

playing tickle monster: Interview with Marqui Hood by phone. 20 Feb. 2013

he'll die on some trip looking for accolades: Interview with Eric Adelberger by telephone. 8 Apr. 2014

Providence was an unusual match: Providence Health & Services and Institute for Systems Biology affiliate to catalyze scientific wellness. Press release. 14 Mar. 2016 https://www.systemsbiology.org/news/2016/03/14/providence-health-services-and-institute-for-systems-biology-affiliate-to-catalyze-scientific-wellness/

Lee's succession plan is death: Interview with Alan Aderem at Center for Infectious Disease Research. 3 Jan. 2013

EPILOGUE: LEGACY OF A SCIENTIFIC ENTREPRENEUR

the tragedy of Lee: Interview with Mark Davis at Stanford University, Palo Alto, CA. 28 Jan. 2013

didn't get the central role in that research that he wanted: Interview with Bill Gates, responses in writing, 25 Oct. 2014

how computer science could approach tough problems of biology: Lazowska, Ed. "Nomination Letter for Fritz J. and Dolores H. Russ Prize." 14 Mar. 2006

one of my sons: Interview with Nathan Myhrvold at Intellectual Ventures in Bellevue, WA. 27 Nov. 2012

to enable people to see what they couldn't see: "Interview with Daniel Kevles by telephone. 3 Nov. 2014

a legacy of having thought properly: Interview with David Baltimore at Caltech. 30 Jan. 2013 *counter to the usual reductionist tendencies:* Interview with Francis Collins by email. 20 Feb. 2013

an enduring theme in his research: Interview with Roger Perlmutter by phone. 20 Aug. 2014

a mechanistic question: Interview with Maynard Olson at the University of Washington. 29 Nov. 2012.

Predictive and preventive medicine will pose striking social, ethical, and legal issues: Hood, Leroy E. "My Life and Adventures Integrating Biology and Technology." A Commemorative Lecture for the 2002 Kyoto Prize in Advanced Technologies. 10 Nov. 2002. https://www.systemsbiology.org/download/2002Kyoto.pdf

why would I want to listen to him?: Interview with James Watson at the Allen Institute's 10th anniversary celebration in Seattle. 27 Sept. 2013

you go to every meeting. You write books: Interview with Ron Davis by phone 4 Feb. 2013

where's the beef?: Interview with David Baltimore at Caltech. 30 Jan. 2013

proteins often behave differently: Shaywitz, David A.; Martin, Joseph B.; Ausiello, Dennis A. "Patient-Oriented Research: Principles and New Approaches to Training." The American Journal of of Medicine. 1 Aug. 2000 https://davidshaywitz.files.wordpress.com/2014/05/por-ajm-2000.pdf